FOR PEOPLE *and* THE PLANET

In the series
Environmental Ethics, Values, and Policy
edited by Holmes Rolston, III

FOR PEOPLE *and* THE PLANET

Holism and Humanism in Environmental Ethics

DON E. MARIETTA, JR.

Temple University Press
Philadelphia

Temple University Press, Philadelphia 19122

Published 1994

⊗ The paper used in this publication meets the minimum requirements of American National Standard for Information Sciences—Permanence of Paper for Printed Library Materials, ANSI Z39.48-1984

Printed in the United States of America

Library of Congress Cataloging-in-Publication Data

Marietta, Don E.
For people and the planet : holism and humanism in environmental ethics / Don E. Marietta.
p. cm. — (Environmental ethics, values, and policy)
Includes bibliographical references and index.
ISBN 1–56639–247–0 (alk. paper) 1–56639–246–2 (case)
1. Environmental ethics. 2. Human ecology. 3. Holism.
I. Title. II. Series.
GE42.M37 1995
179'.1—dc20 94-35667

TO MY WIFE,
Carolyn Gordon Marietta

CONTENTS

FOREWORD

Philosophers since Socrates have insisted that the unexamined life is not worth living, but only more recently have environmental philosophers insisted that life in an unexamined world is not worth living either. Examining his life, Socrates said, "I'm a lover of learning, and trees and country places won't teach me anything, whereas people in the city do" (*Phaedrus* 230d). Socrates loved the city, with its politics and culture; he lived and died defending, and criticizing, Athens. Socrates aspired to the excellence of a noble person, but he hardly knew he lived on a planet. Today we reside on a remarkable Earth; *Homo sapiens* is an Earthling; indeed, we discover—so Don Marietta argues in the pages that follow—that we cannot know who we are without knowing where we are. We need an Earth ethics.

Aristotle placed humans, though they rise up from the humus, the earth, belonging by necessity in social communities, such as villages and towns: "Man is by nature a political animal" (*Politics* 1.2. 1253a). The human genus is animal, but the human *differentia* or essence is to build a *polis*, a town. The human habitat is town or city, which is another way of saying that human life is political and cultural. Humans are the creatures that nature did not specialize but rather equipped with marvelous faculties for culture and craft. Humans rebuild their world through artifact and heritage, agriculture and culture, and political and economic, philosophical and religious, decisions.

Humans have minds coupled to hands; they think about and go to work on their world. They plan what they want; they build themselves a place to inhabit. Deliberately rebuilt environments replace spontaneous wild ones. This is true already in rural communities but becomes especially evident in towns. There must remain a biological understory for our towns as well as for our countrysides, just as the life of the mind requires ongoing biochemistry. But after that, we may think, the values of society are up to us. Our preferences are to be evaluated by culturing those native endowments we call reason and conscience. The town is our niche; we belong there, and no one can achieve full humanity without it. The values that humans cherish in towns cannot be modeled on values in nature, since there are no templates for towns in wild nature. Nature frees humans to do their own thing.

Such a view, emphasizing the exodus that humans make from nature, has been the glory of the last two millennia, first during the classical cultures of antiquity, then during the medieval cultures, and, as though as a culmination, during the scientific, industrial, and technological centuries. Since what we rather overconfidently called our "Enlightenment" we have become modern. The keyword for this phenomenon, in recent discussions, is human "development." But we have not yet found our place.

Indeed, as we turn to the next century, beginning the next millennium, our planet is in crisis. Diverse combinations of nature and culture worked well enough over many millennia, but no more. Development is escalating. We have gained startling powers for the rebuilding and modification, including the degradation, of this home planet. Perhaps the four most critical issues that humans currently face are peace, population, development, and environment. We seem to have reached a critical point in our human self-understanding, facing a future where culture must be reintegrated with nature. The challenge of the next millennium is to contain our developing cultures within the carrying capacity of the larger community of life on Earth.

So we need, as urgently as at any time in our history, indeed more urgently now just because of this long history of separating persons from nature, a discriminating account of the continuities and the discontinuities between persons and their planet, what Marietta calls a "critical holism." We are searching for an environmental ethics, for a satisfactory, situated fitness on Earth. For we humans, too, like the other species, should be a well-adapted fit, even though we no longer achieve that fitness simply by genetics and natural selection, but even more by cultural adaptation and rational evaluation of our roles. How then do we view ourselves in the world?

It is certainly true that humans are a different kind of species. There are some five to ten million species now on Earth; across evolutionary history, over a billion species have come and gone. The procaryotes have speciated in thousands of different lines. The eucaryotes divided into the protists, the fungi, the plants, and the animals. The animals evolved between two and three dozen phyla, with one, the chordates, of special interest. There are some fifty classes of chordates, and one, the mammals, evolved into perhaps twenty orders; and one of these orders, the primates, evolved well over a hundred species. Each of these species is a remarkable biological achievement, from the simplest procaryote (still quite complex) to the most advanced species. The mammals, especially the primates, are amply intelligent. Still, only one of these myriad species achieves enough self-conscious, reflective power to support personality. No other species comes close to producing the transmissible cultures, tens of thousands of them, that characterize the human species.

But does this make us apart from nature? Or are we still a part of nature? If we are a part of nature, is there more to be said about our unique role and destiny? That is the question to which Marietta here turns. We are apart because we can make philosophy, but we find that a philosophy that sets us too far apart is untrue, as well as unsatisfying. Humans are the only species that can be "critical," to use the adjective that Marietta takes from the long-standing philosophical tradition (Greek *kritikos*, a divider, a judge, an evaluator). There are values in the world independently of us, but only we can evaluate this world in overview. Animals live in their own sectors, and so do we, but we also can see outside our sector; we can take a view of the whole. We espouse world views, as can no other species.

Symbolized by the views of Earth from space, what we see is a rounded planet with its community of life, a biosphere, the global location of all our cultures. We are persons on a planet, and we have to keep both nature and culture in focus, like the twin foci of an ellipse, if we are to circumscribe the whole. We humans are at the center of concern, environmental problems are people problems. But our concerns are not well-placed until we see what we are at the center of. That opens for us the opportunity of what Marietta calls a "humanistic holism," one that is also an "ecological holism." As participants in our world, we are not, and ought not to be, free from our world; but we can and ought to be free in it. We are free to be persons on this superb planet, and responsible because we are both caring and free.

An ethic has to be inhabited, as much as an environment. It is a place one lives, part of a niche, which is not just an address but an occupation. Ethics is not just theory but convictions; it is self-involving and life-orienting. A disembodied argument may still be an argument, but a disembodied ethic is nothing at all. An ethic, with its owner, is lived as biography, a story line on location in historical time and geographical space. Our lives are Earthbound, and, with still more particularity, we belong to our native lands, and to lands where we have chosen to dwell.

In one important sense, our ethics should be universal, true at all times and places; still, and equally important, persons dwell in their local cultural and natural contexts, and they constitute their worlds as they live through them. Examining life, Marietta offers us a vision of an ethic that puts persons in their places, fitting there with justice and with love.

Holmes Rolston, III

INTRODUCTION

When I was invited by the planners of an Institute on Religion in an Age of Science conference in 1976 to give a paper on the environment, only a few philosophers were interested in environmental ethics. A conference on philosophical issues related to the environmental crisis had been held at the University of Georgia in 1971 and the conference papers published in 1974.[1] Aldo Leopold's call for a land ethic had been published in 1949, but it was just becoming widely known among philosophers. I first saw it reprinted in a book edited by Paul Shepard and Daniel McKinley in 1969.[2] A few books and journal articles were published in the early 1970s.[3] Some of these were from a religious perspective. A pressing concern at that time was whether Judaism and Christianity were responsible for the attitudes that led to environmental degradation, as claimed by Lynn White, Jr., in 1967.[4] My paper at the conference in 1976 dealt with the effect of religious and humanistic attitudes on treatment of the natural environment. I will return to this issue in Chapter One.

From a small beginning in the late 1960s, environmental ethics has grown within not quite two decades to be a recognized specialty in philosophy and a matter of interest to people in a wide range of academic disciplines and environmentally oriented professions. The concern about the natural environment and the proper behavior toward it is not just an academic and professional interest. There is now a goodly host of civic activists who want the society to take serious action to preserve the natural environment. Throughout the nation people are recycling aluminum, paper, and glass. Service stations and garages are saving used oil and using gasoline pumps that lower the amount of vapor released into the atmosphere. People are refusing to buy goods with unnecessary packaging or in containers that cannot be recycled. Businesses and institutions are buying products made from recycled materials. Private groups, cities, counties, and states are buying undeveloped lands for preservation. National parks are used in part for preservation of wilderness. These things are just part of a growing environmental movement.

People are beginning to think about the natural environment more than they used to, and they think about it differently. Old notions about the environment are giving place to very different beliefs, and new attitudes are becoming widespread.

For roughly two thousand years, the mainstream of Western thought minimized the importance of the natural environment, except as it was useful to people as farm or ranch land or as natural resources. The human person was looked upon as separate and different from the rest of living things and the lands and waters that sustained them. Until this century very few people thought of humans as a part of nature. The "man"-apart-from-nature view was dominant. I use the term *man* deliberately because it was man, the human male, who was thought to be apart from nature, as I explain further in Chapter One. This dominant concept of human, especially male, superiority to all of nature shaped the thinking of all but a handful of the influental thinkers. Plutarch was one of the few Roman intellectuals who thought people were morally bound to treat other animals with kindness; most of the classical thinkers after the time of Socrates stressed the difference between humans and nature. In the Middle Ages Saint Francis loved nature and even called the creatures his brothers, but he was not a typical thinker of his time. The early Renaissance humanist Petrarca wrote of his appreciation of the beauties of nature and the inspiration he received from climbing a wild mountain side, but his was an unusual sentiment that few other humanists adopted. Giordano Bruno taught that all nature is one, but this unorthodox view disturbed the guardians of theological purity and contributed to his being burned at the stake. Spinoza's god was nature, the one nature of which all things material and spiritual are manifestations, but this was a radical departure from Jewish and Christian theism, and few people could accept it. The prevailing view of the natural world in Western thinking alienated human beings from nature. For most people, nature was a hostile or indifferent force that worked against the welfare of humans and their intellectual and spiritual development. Nature was available for humans to use, but nature had to be conquered to be useful. It was human toil that turned the wild places into gardens. People felt little kinship with the rest of nature.

The dominant conceptual model of these twenty-some centuries did more than separate man from nature. It tended to separate all things from every other thing. The "man"-apart-from-nature paradigm was atomistic rather than holistic. I call it atomistic because it was a way of seeing all things as separate atoms, which were only coincidentally parts of other things. This thinking stressed the individuality of things and minimized the connections between things. The differences between things became their defining characteristics, while the qualities that things had in common were depreciated. This affected the common view of what humans are and of what the world is. One of the significant effects of this atomistic way of thinking can be seen in ethics, which stressed individual duties and individual rights and claims against other people, with little or no sense of collective responsibility toward the earth.

The individualistic, atomistic way of conceiving of things made the nature of each thing internal to it, but the relations that a thing had with other things had to be recognized. The relations that a thing had with objects that were not a part of itself were thought about hierarchically, with each thing ranked in relation to other things. The things with which a thing was related were either better or worse than it, above or below it, more or less important. People were ranked from ruler to slave. Aspects of nature and of human life were ranked, one above another. Soul was above body, reason was above emotion, human was above animal, and the European was above all other cultures and peoples. The most pervasive ranking of all, as a result of which half the human race was affected adversely, was the placing of man above woman. The rankings of human above nature and man above woman seem to have had the most profound affects on human behavior. The opposition to nature and the sexual chauvinism that saw the human male as the normative human being were merged in a dominant paradigm accepted by virtually all philosophers and religionists. Feminist philosophers rightly see a close connection between the oppression of women and the abuse of the natural environment in Western culture.

The separation of humans from nature paralleled a division of body and mind within each person. Psychophysical dualism, which saw the soul or mental aspect of human beings as the real person, with the body as a temporary housing of the soul, diminished the importance of the natural environment. This dualism also affected the position of women, with intellectual life becoming the domain of the male and caring for the body the service of women.

The amalgam of these several ideas made a powerful and comprehensive philosophical foundation for medieval and modern culture. In Chapter One I will explain in more detail the "man"-apart-from-nature view and contrast it with the new holistic models that are replacing it in the thinking of many people.

The obvious danger in which humans have put themselves by abusing the natural environment, shown by biological studies and the new science of ecology, has made increasing numbers of people take a critical look at the old ways of thinking about nature and look toward a new paradigm that sees human involvement in the biosphere in a more constructive way. Recognizing that the trashing of the planet was more than an aesthetic offense has made people rethink their place in the natural world. Philosophers have begun to question the adequacy of atomistic, hierarchical, and dualistic thinking, and people in all walks of life have become aware of the inadequacy of old ethical principles in meeting the challenges of human damage to the planet on which life depends. At the same time feminism has forced a rethinking of basic values and standards of human life.

In Chapter Two I will explore a new perspective on the person as a part of nature. The new paradigms that are replacing the apart-from-nature view employ several images to make the relationship of people and the natural environment more vivid. A community model sees the human person as a citizen of the biosphere. An organistic model sees the human person as an organ in a larger organism, the planet Earth. Earth has been pictured as a lifeboat and as a spaceship. Each of these images has implications that will be examined; each of them has its strengths and its limitations as a way to visualize the role of humans in the world. The many implications of seeing the human person as a part of nature need to be identified and examined.

A major task for philosophers is helping to develop an adequate system of ethics for living in harmony with the natural environment. At first there was little realization that a new approach to ethics was needed. Many ethicists thought that an extension of traditional moral principles to cover issues raised by deterioration of the natural environment would be adequate. One of the first moral issues of which they became aware was related to the finitude of natural resources and the fact that a few nations were using up most of the nonrenewable resources. The main questions faced at first dealt with just distribution of resources among nations and socioeconomic groups, and they involved no new moral principles. Extending the principles of justice and utility to include animals and unborn generations required some refinements in understanding such notions as the concept of the moral community, so that the moral standing or considerability of nonhuman animals and future generations of humans could be recognized. Including plants in the range of moral concern extended traditional moral principles even further. This new development, extending ethical principles to cover all living things in the range of moral concern, is known as biocentrism or "second-phase extensionism."[5]

For nearly two decades now, environmentalist moral philosophers have been attempting to develop a new kind of ethics, which we can call ecological ethics. The new approaches to environmental ethics are often called environmental holism because they are far more inclusive than the traditional ethical systems and recognize that the human person is a part of the natural system that supports life on this planet. Some of these new approaches to ethics include nonhuman animals in the range of ethical concerns. Others, often referred to as biocentric approaches, include all living things. Ecocentric approaches stress the value of the planet as a whole and uphold moral responsibility toward the whole system of nature. The ethic that I support is holistic because it stresses the moral importance of the interrelationships of animal and plant species and their common dependence on the ecosphere, the whole system of living things and the water and minerals on which life depends. I will call the holism that I advocate "critical holism." This is done partly for convenience, to avoid having to repeat such phrases as "the type of holism I advocate," but there are two senses of the word *critical* that are relevant to that particular kind of holism. First, the sense of importance; my holistic viewpoint recognizes the critical significance of understanding that humans are a part of nature. How they are part of nature will be treated later. The second sense is that of discrimination, of close and careful examination. Critical holism attempts to avoid an exaggerated or fanciful concept of the unity of nature. It tries to avoid going very far beyond actual findings of ecology. There are approaches to ecological science that do not focus on the holistic aspects of natural systems. In calling my approach ecological, I do not claim that it is supported by all points of view in ecology.

Holism has meant various things to different thinkers, and I do not think that some holistic views are sound. In calling my view holistic, I do not mean to place my ethics in the context of extreme versions of holism that neglect or minimize the role of human culture and the value traditionally given to the human person. The ethical approach that I describe and advocate here does not sacrifice to concern for the natural environment the traditional concerns of humanistic ethics, such as justice, human welfare, the advancement of knowledge, and the avoidance or mitigation of needless pain. For this reason, I also refer to critical holism as humanistic holism.

More detailed description of the concept of critical holism and the nature of and limitations on the interrelationships of humans and other things in nature will be examined later, especially in Chapter Two.

Upon the basis of critical holism, I defend a humanistic ethic that calls for a duty more inclusive than duty just to humans, or even just to living things; it recognizes a duty to be responsible toward the nonliving things without which there could be no life. The obligations to the whole system of nature do not supplant duties to other persons and to other living things; the duties to the ecosphere supplement the other duties. Responsibly meeting duties of these different types requires new approaches to ethics that I will discuss in the several chapters of the book.

Environmental holism and the concept of duties to the ecosphere has been disturbing to a number of thinkers. The ecosphere, whether it is thought of as a community or as an organism, has implications that have troubled a number of philosophers. If the welfare of the whole community is given priority, the human individual may no longer be the focus of moral concern, with significant loss of freedom and moral rights. This has been recognized as a danger when only the political community was the larger whole to which the individual was subservient; with the whole planet as the community, the subordination of the individual might be even more devastating. Talking about nature as an organism raises even greater fears than the concept of nature as a community; an organ might have less moral significance than even a lowly citizen. The principles of humanistic ethics, which governed the relationships and interactions between people for thousands of years, might be overwhelmed by duties to the whole natural world.

In Chapter Three this concern over the loss of individual rights and freedom will be examined. The danger that humanistic ethical concerns will be sacrificed to survival strategies remains a serious threat, which I do not take lightly, but I argue that it is not environmental holism itself that we need to be concerned about. A critical holism that avoids the pitfall of being too abstract in the way it pictures the role of humans in the natural world and the pitfall of being too reductionistic in its understanding of what the human being is does not pose a danger to humanistic ethics. It is excessive abstractness, which loses sight of what is actually happening in the ecosystems, and reduction, which ignores morally significant qualities of humans in general and of the individual person, that threaten the value of the individual and the ethical standards that have fostered peace and justice among people and nations. A properly understood holism does not entail draconian measures that sacrifice the moral significance of individual persons and human relationships. An ethical system that recognizes a number of sources of moral insight and a broad range of moral concerns will enable the reconciliation of concern for the natural environment, concern for animals, and concern for justice and human welfare.[6]

Many environmental philosophers have played down talk of concern for humans, trying to avoid a biased approach to environmental matters that assumes the moral superiority of human beings and a special importance for human interests. They have seen that assumption of human superiority to everything else in the world has been similar to notions of racial or national superiority. They believe it was this focus on humanity that led to attitudes that were destructive to the environment. In Chapter Four I will review the development of the rejection of what is termed *anthropocentrism*. I will point out the good that has resulted from the realization that Earth was not made just for humans and that morality should not be limited to narrow human concerns. I have come to believe, however, that even though the deep-ecology and biocentric approaches to ethics, which claimed that humans are not morally more significant than other living beings, have had a beneficial effect upon moral philosophy, the attack on anthropocentrism should be reexamined. The positions in environmental ethics that challenged the limiting of ethical concern to human interests woke many of us from deep dogmatic slumbers, but blaming all the harm done to the natural world on anthropocentrism may have been a mistake. I examine the possibility that concern for humanity might offer a better approach to environmental ethics than has been recognized by many environmental philosophers. Of course, a human-centered ethic must not be shortsighted and must take account of ecological knowledge to be an adequate ethical stance. I believe this includes the adoption of a moderate holistic view of humanity that recognizes that humans are truly a part of nature, even if being part of nature is not all that humans are. Only a holistic anthropocentrism can be a scientifically credible and morally adequate humanism.

A new approach to ethics, forged in response to the crisis of environmental destruction, faces the task of showing that it is justified philosophically. The new approach to ethics requires a new understanding of what ethics is and how it works. There is more to philosophical ethics than the making of judgments about which acts are right and which wrong. Understanding the nature of ethics has always required attention to the meaning of ethical terms. More recently it has involved a more extensive analysis of ethical language and the nature of ethical judgments. One of the most difficult tasks of the moral philosopher is addressing the logical issue of justifying moral principles. Ecological ethics must try to justify the new principles that underlie the new duties to the ecosphere. It must try to explain the reasoning that underlies the recognition of the newly realized duties. The new environmental ethics must be shown to be logically coherent, consistent with our knowledge of the natural world, and able to give moral direction as we face difficult environmental questions. Explaining why a moral principle is logically justified is a difficult but necessary aspect of moral philosophy. I believe it is morally wrong to destroy an estuary just to dig a channel for coal barges so someone can make more money than would be made if already existing ports are used. Some people do not agree with this. How can I explain why I am right and they are wrong? This can get complicated, but it needs to be attended to. The logical basis of the new environmental ethic must be made clear and shown to be sound.

At first moral philosophers tried to justify environmental ethical principles in the same ways older moral principles had been defended. A growing number of environmental philosophers came to believe that the new ethics raised not only new moral issues but also called for new moral principles to be informed by ecological knowledge. This raised new logical issues regarding the defense of these new moral principles. There must be some explanation of how moral knowledge (knowledge of what is good and what is bad) can be related to factual knowledge, including knowledge about the natural environment. For a long time moral philosophers have realized a need to avoid what is called the *is/ought* impasse, which results when attempts are made to use deductive or inductive reasoning to explain the relationship between factual information and moral evaluation. I will explore an approach that does not attempt to reach moral knowledge through deductive arguments. I will base this approach on current views about perception and the formation of beliefs. An important part of this is seeing the role played in moral judgment by a person's individual world view. World views are subject to criticism, and this enables the determination of which world views, along with the moral judgments that are made in their context, are adequate and which are inadequate.

I also explain the movement from moral knowledge of the good and bad to a sense of moral obligation to do what is good. Some moral philosophers have thought that a logical connection must be found between knowledge that certain things are good and others bad and an obligation to do what is good. My approach, based on an examination of actual moral experience, finds no gap between knowledge and obligation. I find that recognition of goodness and badness is accompanied by recognition of a fittingness between what is seen as good in a given situation and certain behaviors that are appropriate to fostering the good and avoiding the bad. This sense of fittingness shows which behaviors one should do and which acts one should refrain from doing. With this recognition of fittingness comes a tendency to seek what is good and avoid what is bad. The significance and the limitations of this connection between knowledge and a sense of obligation will be examined.

A number of logical issues will be explored, especially in Chapters Five and Six. These chapters can be skipped or saved for a later time without loss of continuity with the other chapters.

Much of the new environmental ethics has been based on the concept of value in nature. Some of the ethical approaches are grounded in the attribution of intrinsic value to animals, to the biosphere, or to the whole ecosphere. This has raised a number of difficult questions, such as questions about the objectivity of values. Whether values can be objective and thereby able to give a sound basis for ethics will be examined in Chapter Seven. An effort to provide a clear taxonomy of value terms will be followed by an analysis that supports the conclusion that values are not properly understood as being either objective or subjective. I will examine the ethical significance of abandoning talk of objectivity with respect to values and argue that there will be no loss to ethics. There are clearer and more accurate ways of saying what we have tried to say in talk about objectivity, and using more descriptive ways of talking will not rob us of value, but may give us a better grasp of the values that are significant to us morally.

The obligations of which we become aware in seeing the fittingness of certain behaviors, including the fittingness of obligations toward the natural environment, are significant within specific situations. The meaning of the situation for the person who will act within the situation provides part of the context within which a behavior is fitting. It is within the context of *this* situation that *this* behavior is seen to be the right thing to do. An analysis of actual moral experience does not support the notion of moral absolutes that are binding upon all rational creatures regardless of the situation in which they must act. It supports the idea of contextual obligations, have meaning and are morally demanding because of a context that includes the person's individual world view. This world view incorporates the person's knowledge and beliefs and values in a comprehension of the world that includes cognitive, valuational, and volitional aspects. In Chapter Eight I will explain and defend the contextualist approach to ethics. This chapter concludes with an explanation of how holistic contextual ethics can contribute to the resolution of difficult and confusing moral issues, the hard cases.

In Chapter Nine I will explore moral pluralism, which is much more controversial than contextualism. Contextualism has been a part of moral theory almost from the beginning of ethics in Greek philosophy, even though some ethical systems tried to overcome its role in morality.

Pluralism is more threatening to traditional moralists because this approach uses different moral principles that cannot be reduced to one basic principle and cannot be logically derived from one master principle. I will survey some of the recent literature on pluralism and explain how use of different principles is justified. Moral judgments can be shown to be of very different kinds, such that different principles are in fact doing somewhat different jobs in the several cases, such as deciding what one should do, judging past behaviors, judging moral character, and making social policy. The situations in which moral judgments are made can vary from those involving people who are well known and whose lives one can affect very deeply to those involving people who are quite distant and who can be affected only slightly. Decisions involving wild animals and future generations are certainly not like those involving one's family or close associates. Different moral principles may very well be more appropriate in one sort of situation than in another. I will argue that a pluralistic approach is justified if a clear explanation can be given of how the kind of principle followed in a given case is justified by the particular sort of judgment that must be made and features of the context in which the judgment is made.

In Chapter Ten I will explore the causes of moral disagreements and the role of disagreement in moral discourse. I will argue that moral consensus is not necessary; moral disagreement can have the positive effect of showing alternatives and fostering creative approaches to moral matters. Moral disagreement does not call for the subjectivist view that all moral positions are equally valid, nor does it justify moral indifference. Environmentalists need not question their moral insights because there are people who deny moral responsibility toward the planet. In this chapter I will argue that a contextualist and pluralistic ethical system allows desirable flexibility and creativity in finding the appropriate and responsible moral behavior in new and complex situations. It does not allow justification of whatever a person might happen to fancy. Moral beliefs are not all equally justified. Previous chapters have explored the ways in which moral beliefs can be evaluated.

Equally honest and well intentioned people with equally good scientific knowledge can disagree, so disturbing disagreements about ethical treatment of the natural environment will occur, and this should not be surprising. Moral judgments are subject to varying degrees of uncertainty. There will be times when we must act upon probability. Some moral schools have claimed to give complete moral certainty, but this claim has not been generally convincing. I argue that the notion that moral systems should provide certainty has had a destructive effect upon morality because this claim leads to unreasonable expectations of moral systems, leading to skepticism and subjectivism when the moral philosopher fails to produce the promised certainty.

In all areas of life we face uncertainty and have to act upon probability. Morality is no exception to this. Uncertainty does not destroy moral knowledge, just as uncertainty in other fields does not destroy the difference between well-founded beliefs and beliefs that are not credible. In all areas of life we must act on probability, but that does not obscure the difference between good decisions and bad decisions.

In the final chapter, I will defend the moral adequacy of the version of humanistic holism explained in this book. This ethical approach provides moral demands that are binding upon the agent. They are not absolute in some abstract way; they derive their meaning and significance from a context that includes the person's individual world view. This does not, however, keep the moral demands from meeting the standards that moral philosophers should expect of moral claims. They transcend the desires and self-interest of the agent and can be universalized in a significant sense. A contextualist and pluralistic ethical approach that acknowledges the element of uncertainty in moral judgment will help overcome a practical problem in moral philosophy, a problem caused in large part by a mistake of moral philosophers. The problem is a largely unexamined subjectivism that leads many people to consider ethics a matter of personal opinion, a matter of whim and fancy. This naive notion even parades as the modern person's sophisticated liberation from dogmas and superstitions of the past. Why has this subjectivism in ethics taken such a hold on the thinking of people who are far more demanding in regard to other areas of study? The fault, I argue, is the mistaken quest for certainty in ethics, which has led philosophers to favor abstract and reductionistic ethical systems that provided a spurious impression of clarity. Ethical theories that make valuable contributions to our moral understanding and to our making of ethical judgments were harshly criticised because they did not provide indubitable abstract principles. Moral philosophers themselves inadvertently taught other educated people to think that moral philosophers had not achieved very much. The tragedy in this is that many people are unaware of the help they can receive from philosophical ethics.

I try to show that a scientifically credible humanistic holism, which reconciles concern for the environment with more traditional humanistic ethical concerns, such as freedom and justice and human welfare, is workable and is the best approach to ethical concern available to us. This approach does not provide absolute certainty because it is contextual and pluralistic and must resolve conflicts between duties to individual persons, to society, and to the natural world. Humanistic holism recognizes the complexity and uncertainty of life, and it accepts the need to consider everything that is relevant to the making of ethical decisions and to act in good faith on the basis of probability. While we live complicated and uncertain lives in a world we still do not understand completely, we do well to accept the goal of ethics as providing the best guidance now available to well-meaning people. That we will make some mistakes in morals, as we will do in other areas of life, must not make us reject a moral theory that is the best available to us. We can live responsibly, doing the best we can to create a sustainable society within a natural environment that can continue to support life. Even though it may not be easy or simple, we can live according to a humanistic and holistic ethic, one that seeks the good for people and for the planet.

CHAPTER ONE

Changing Perspectives on Nature

Natural Resources vs. Values in Nature

During the past few decades, the way knowledgeable people think about the place of humans in nature has changed. The way people conceive of nature itself is undergoing a change. This is a change in the fundamental conceptual model in terms of which people relate to the world, the interest they take in it, and the value they see in it. This change in people's conception of the world is of tremendous importance. It affects people's daily lives, local and national politics, and economic activities. One type of conceptual model, the scientific paradigm, is more basic than hypotheses and theories. The paradigm is the intellectual framework within which a theory has its meaning. It is the basic mental picture within which an hypothesis is important enough to be tested in experimentation. One of the most important changes in the history of science was the rejection of the Ptolemaic model of the solar system, with its picture of the earth as the unmoving center, and the acceptance of the Copernican system of a central sun with Earth and other planets in orbits around it. The change in conceptual models made some old concepts and theories meaningless. Speculation about epicycles, the notion of perfect spheres, and other matters that were important within the old model were of no interest within the new model. Some questions may remain important through a paradigm change because the old and new models may not be totally different.

Many conceptual models are not as clear or as closely related to observational data as are scientific paradigms. Some models loosely incorporate parts of several scientific models along with social and psychological concepts. A conceptual model can be poetic, even fanciful. The "man"-apart-from-nature concept is a model of the relationship of humans to the natural world. This conceptual model made sense of some ideas that many people now find unacceptable. Many of us have a different conceptual model, one changed from the apart-from-nature model to a person-as-a-part-of-nature model. This change in conceptual models makes some old questions meaningless and some new questions consequential.

The dominant view of nature in Western society for more than twenty centuries has been that nature exists to serve human beings. The human use of nature has sometimes been wanton and wasteful. At other times nature has been used carefully, with an eye to the future. Since the science of ecology has taught us about the fragility of natural systems, a restrained use of the resources of forests, croplands, mines, and oceans, usually referred to as conservation, has been practiced more frequently. Conservation has not been universally practiced, however, and the forests and soils of many countries have been greatly diminished, while the waters and the air have been seriously polluted. Even though some people carefully husbanded the resources of nature and other people used these resources prodigally, almost everyone had the same basic notion of the value of nature. The basic view of nature held by people who were unhappy about the losses and the pollution, those who practiced conservation, was often the same as the view of the wasters and defilers; nature was seen as having value only as natural resources for humans. Arne Naess calls the conservation approach "shallow" ecology, which he described as the fight against resource depletion and pollution for the benefit of humans.[1]

The concept of nature as natural resources and the belief that humans are the rightful heirs of nature's treasures is indicated in many prevalent attitudes. Many people view as wasted any land left natural, any stream not used by someone, and trees left to decay in the forest. Many people simply cannot understand why people are concerned about the spotted owl and protection of the old growth forests. Tree farms seem to them better and more useful than old forests. With tree farms there can be easy access to just the right kinds of trees, that is, the species of tree desired for timber. The concept of nature as resources for humans is seen in our basic economic theory, known technically as classical liberalism. This economic approach, which underlies the thinking of capitalism and socialism, even Marxism, looks to the exploitation of the natural environment to provide for the satisfaction of human wants. The economic philosophy of John Locke expressed clearly the concept of nature as resources for human use. Locke held that the products of nature have no value until they are claimed and used by someone, improved or made into something useful. Locke helped to make a labor theory of value a basic feature of the economic theories underlying the European and American economies. According to Locke, nature contributes only a very small fraction of economic value; human effort and use make land, streams, and plants valuable. The dominant economic principles of Western industrial societies recognize little value in nature itself; the goal of economic activity is held to be the satisfaction of human desires. An efficient economy is one that provides for meeting the interests of human beings to the greatest possible extent. This focus on human benefits is implied by the ideal goal of economy, Pareto optimality, which calls for a society in which no one can be benefited without taking something away from another person. This optimal condition would hold when natural resources are used with utmost efficiency. There is no place in these theories to consider the environment apart from human interest in it. One way to see this is in cost/benefit analysis, the widely used method for determining whether a project should be undertaken. Cost/benefit analysis, as it is commonly used, can take into account only the economic effect of an action or policy upon human beings. No value of unexploited nature can be recognized; nature can be assigned no value except on the basis of human interest in some aspect of nature. The only way to view nature is as a resource for human betterment. The only human satisfactions that can be measured and enumerated are, of course, those that can be seen in economic terms. Aesthetic values, spiritual values, and other values that a person might find important tend to be ignored in the making of public policy, which is usually based on cost/benefit analysis. Costs and benefits must have a market value to be included in the assessment of proposed projects.[2] There has been some objection to the basing of public decisions on market-place values,[3] and recently some economists have acknowledged that market value concepts apply to "economic man," which is a useful model for economists. Others have recognized the limitations of cost/benefit analysis and advocated giving a "shadow price" to values not subject to market transactions.[4] Some people object to shadow prices, which attempt to put a price on something that is cheapened by being given a monetary worth. Just as many people object to putting a price on a human life, some people object to considering the beauty of a forest or a river a thing that might be exchanged for money. The concept of things being priceless is still meaningful to many people. The problem comes when, in the making of public policy, the things that are priceless are treated as things of no worth. The worth of platforms for off-shore oil exploration can be established, or at least estimated, while the value of reefs that are not contaminated by oil is treated as something which exists only in people's minds.

Until recently very few people gave much thought to the limited supply of many resources. Even now many people ignore the finiteness of nonrenewable resources, such as petroleum, and the danger of using renewable resources, such as timber, faster than they can be regrown. Some people express great confidence that suitable substitutes for oil and other scarce things will be found in time. Man has always found an answer, they say. Pete Gunter calls the Western belief that nature is infinite as natural resources, while man is infinite in his potential for conquering nature and solving all problems, the man infinite/nature infinite myth. This man infinite/nature infinite approach still gives people unrealistic expectations and blinds them to potential catastrophes. Gunter shows the connection between this myth and a view of nature that has no realization of the complex interdependency within natural systems.[5]

Even some people who have consciously rejected the soul/body dualism that allowed them to think of themselves as souls separated from nature still fail to see the extent to which humans are a part of a fragile natural system. Once people become aware of the relationships between living things in the complex system of nature, including the relationships between humans and other living things, they can see that the concept of humans as radically different from all other living things is a myth. It is a myth difficult to take seriously if one is aware of the human dependency on interactions with the other participants in the natural scheme of things. I remember the effect on my thinking when I learned about the multitude of organisms living inside me and on my skin, enabling me to live a healthy life. I have realized the advantage of having a yard crawling with lizards and aflutter with birds, with plenty of worms and smaller things in the soil. Where would any of us be without the good air to breathe, coming to us courtesy of the plant life on the land and in the sea? A holistic view of the place of humans in the natural system challenges the notion that humans are the only heirs of all the natural wealth by showing that this wealth serves organisms of all types. The environment that makes life possible for humans also enables other organisms, great and small, to live, and we can see that our human lives are dependent on those other lives. It is difficult to maintain, in the face of a holistic understanding of our place in nature, the old view that this all exists simply for us. I have enjoyed seeing a tortoise eating a prickly pear, a fruit that hardly seemed designed to meet my needs. I have eaten prickly pears, but not with the ease with which Brother Tortoise munched it, and not with an enjoyment that would make me seek them out except as survival food. I think it is good for us to see the other animals hunt, eat, and go about their lives. The natural environment becomes before our eyes a wondrous system of vast creativity, and we come to realize that human exploitation of this fountain of life is not its only value. The issue of values in nature is a topic of profound importance that I will treat in Chapter Seven.

Separations vs. Interconnections

The resource-use approach to nature is not the only view that people have had of nature. It seems that early Mediterranean cultures had a nature-centered paganism that recognized that people were part of nature, along with all other life forms. The Hebrew scriptures provide evidence of paganism in the Near East; the writings of the prophets from about the ninth century voice condemnation of the religious practices of indigenous people and the influence of those practices among the Hebrew people. These pagan practices gave expression to the involvement of humans in nature, especially in the changing seasons, the growing season with its promise of security and well-being, and the season when nothing could grow and life was threatened.

Across the Mediterranean Sea, the Ionian culture was not marked by mind/body dualism, and the people of the early Minoan culture of Crete (and later the Mycenaean culture of the Greek mainland) worshipped gods associated with nature. The Cretan murals found at Knossos indicate delight in the beauty of nature, with birds and animals, including the dolphin, providing the subjects of many paintings.

On both sides of the Mediterranean, mystery cults, some lasting into the Roman era, were concerned with the annual cycle of fertile times, the warm season in Europe and the wet season in Egypt, and times when nature seemed to be dead.

The familiar story of Demeter and Persephone is the myth of the highly respected Eleusianian mysteries. Other myths related the stories of the consorts of goddesses who were associated with nature; the consorts were mutilated or killed and later brought back to life. These deaths and resurrections were symbols of the dreaded time when the crops did not grow and the return of the season of growth. The pagan religions show awareness that human welfare could not be separated from the natural world. The pagans seemed aware that they were a part of nature, and it paid to be on the right side of nature.

The early European and Near Eastern attitudes toward nature are not unique. Some American Indian cultures had beliefs and myths that affirmed the unity of all things in nature. There was no mind/body dualism in the beliefs of these cultures; there is talk of spirits, but a careful study of these cultures shows that spirit was associated with life, even the life of non-human animals, plants, places, and elements of weather, such as winds. This way of understanding the place of humans in nature is seen clearly in John G. Neihardt's book, *Black Elk Speaks*, which tells the life story and the visions of an Oglala Sioux holy man. First published in 1932, this book reveals a concept of human life radically different from the perspective of man-apart-from-nature.

In Europe, the nature oriented mythologies gave way to religious philosophies that were concerned with the salvation of human souls, apart from the natural world. I will discuss some of the sources of this change in Chapter Two. Now I want to point out that the separation of humans from nature was not the only separation that marked the later western European approach. Mind or soul was separated from the body. A dualism of mind and body came to dominate the thinking of western culture, and this fostered a depreciation of the body and of the physical world because the most important part of the human was now thought to be apart from nature. Nature was hated and feared or simply held to be of little importance. The theistic view of nature in Christianity and Judaism was not derived wholly from the Hebrew scriptures, some passages of which express a great appreciation of nature. Judaism and Christianity were influenced by popular Greek philosophical ideas current in the Roman period, which were the source of attitudes that varied from hostility toward nature, seen as the source of evil, to a disdain for nature, seen as unable to provide salvation and often tempting the soul away from its proper object of love, the transcendent god, and from its proper concern, the everlasting life that follows the mortal life. The early pagan religions, which considered the human as part of nature, were bitterly condemned by most theists of the early days of Christianity, along with the music, art, and philosophy of pagan culture. More tolerant theists in the late Roman empire, such as Augustine, did not refuse to use the contributions of pagan culture that they deemed harmless and compatible with theistic religion, but the paganism with which third- and fourth-century Christians were most familiar was the highly intellectual Neoplatonism, which interpreted the pagan myths as symbols of the soul's quest for escape from the body in order to secure for the soul reunion with god. Christians and Jews who borrowed too much from this intellectual pagan philosophy were eventually condemned as unorthodox; their interpretations stressed too much the mystical union of the soul and god and the "pantheism" to which some forms of mysticism were prone. The robust paganism of earlier days, which saw divinity in the growing of crops and in other aspects of the natural life, was dead except in remote rural areas, the places where "villains" and "heathens" lived. For civilized people, salvation was to be found apart from nature, in renouncing the world, the flesh, and the devil.

The attitude of people toward the wilderness, the place of nature "in the wild," shows a great deal about the Western attitude. The Hebrew culture, as seen in the Bible, seems always to have had a negative attitude toward wilderness, sometimes seen as a desolate place of suffering and exile. The New England Puritans accepted the harshest biblical view of wilderness and associated wild areas with the devil; people who visited the forests were considered witches who consort with the devil. The stories of Nathaniel Hawthorne give an interesting picture of this attitude.[6]

Until recently, most people thought of wild areas as ugly, dangerous, and a wasted resource. Land was of little value until it was cleared. Gardens and parks were held in high regard, but wilderness was at best wasteland, at worst a blight to be removed. The wilderness, nature untamed by people, had nothing to offer the human mind and spirit. Salvation did not need nature, but an overcoming of nature.

This disdain for nature is still to be found in contemporary thinking. It was ironic to hear former Secretary of the Interior James Watt put down concern about the natural environment on the ground that salvation would not be of this world. The irony is that the official protector of the environment expressed such a disregard for it. The connection between Watt's other-worldly theology and a lack of concern for the forests and rivers is not surprising. Typically such a theology considers the human soul the real person and the body a temporary housing of the soul. Historically and in the present, a close connection can be seen between the separation of soul and body within the person and the separation of people from their natural environment. Psychophysical dualism does not lend itself to concern for the natural world, but it is not only because of its danger to the wild places we love that we should oppose the mind/body dualism.

There are a number of problems with a dualistic view of the human person. It is very questionable psychology, which no longer has the widespread support among psychologists it once had. Research in physiology and psychology does not support the view of the human person as a mind or soul that is independent of the body and inaccessible to other people unless the person chooses to reveal innermost secrets. From a philosophical perspective, since the beginning of modern philosophy, conceptual difficulties in soul/body dualism have been recognized. There has been no clear explanation of what such a soul could be and how it could function independently of the body. Contemporary philosophical supporters of dualism are no longer defending the concept of the self as two distinct substances. Their concern is that mind not be explained as merely an epiphenomenon of brain processes, that its reasoning processes and its normative judgments not be reduced to chemical activity of the brain. Among philosophers, there are few defenders of the notion of the soul as an entity separate from the body.

The notion of human separation from the natural environment receives no support from biology and ecology, which have clearly described human dependence on the biosphere. Studies of evolution show the kinship between humans and other animals. Dualism and the separation of humans from nature must be replaced with a holistic understanding of human beings and the natural world. What this holism should be, the nature of human involvement in nature, and the importance of human culture will be treated in some detail later.

Hierarchical Thinking vs. Organic Systems Thinking

Related to the dualism of the old way of thinking was another element, atomistic thinking, by which I mean thinking in terms of strong differentiation of each entity from all others, seeing each thing as a separate "atom." Connections between things are treated in this perspective as accidental and mechanical, not affecting the essential nature of the thing itself. In the way of thinking about nature that has been dominant for two millennia, there was no understanding of the complex interactions on which life depends. The land was not respected as the foundation of all life; it was not even thought of as a whole, the land. Land was this field and that lot, all of them market commodities. It was a way of thinking about individual things, separate things, with little attention paid to the relationships that made nature a vast life-supporting system. The atomistic pattern of thought focused on individual things, rather than on the wholes within which individual things have purpose and meaning. Killing an animal was just killing one animal, and what is one animal more or less to a person who does not think of species and of roles in an ecosystem. It was always a matter of one tree, one stream, one waste dump, one field cleared, one mine dug. Even when it was noticed that game was less plentiful or the field was producing less corn, there was little thought of connections between things.

Atomistic thinking affected the way things were defined. The most distinctive quality of a thing was taken to be its defining characteristic. The definition of what humans are suffered from this tendency; I say "suffered" because it did involve some real loss of fullness and richness in the understanding of what it is to be human. "Man is the rational animal" may sound innocent enough until we see how we have neglected the fact that human beings are much more than intellects. We humans are emotional animals, lustful animals, and aspiring animals, who have other characteristics that shape and direct our behavior; thinking of ourselves as beings who should place most of our concern on our intellects is to deny a great part of who we are. Interestingly, Aristotle, who seems to have been the first thinker to deal carefully with the matter of taxonomy, did not think that the definition of the human as the rational animal was a complete definition. He held that the human is fully defined only in a description that includes the qualities shared with other species.

We failed to avail ourselves of valuable parts of our philosophical heritage: Aristotle's realization that a species is defined by every class in which it has membership, Leibniz's view of a harmonious cosmos in which external relations were recognized as definitive aspects of each entity, and Spinoza's concept of all things as parts of one divine substance. Instead of building on such rich insights, we fashioned our common understanding of ourselves and other things in an excessively individualistic and isolating way. Thinking in this way, we had no sense of belonging to the natural order of things, no sense of kinship with our fellow creatures in the world. Many people had little sense of kinship with other humans.

The ethical systems that regulated human relationships were affected by atomistic thinking, and in turn the ethical systems strengthened the differentation and separation. Modern ethics has been mainly a matter of individual responsibilities and individual rights. Separate persons had duties toward other persons; these duties were usually correlated with rights that individual persons claimed as individuals. Groups of people were usually seen as collections of individuals. Social issues were awkward problems in such a framework. When no one specific person was uniquely responsible, social good did not fit the ethical pattern. To hold corporations or governmental units to be responsible for something was a puzzling matter for atomistic thinkers. A kind of cult of individualism gave added strength to separations between person and person and between person and world. The rugged individual who had no need of anyone, who did not care what anyone thought, was the hero. Success was won in fierce competition; cooperation was often seen as a concession to weakness.

Without a strong sense of group responsibility for the natural environment, individual persons seldom considered the care of the land their responsibilty, and the ecosystem suffered. Even when a problem was noticed, the prevailing attitude was "it's not my job."

The exaggerated individualism of the apart-from-nature model was combined with another way of thinking that some environmentalists, especially ecofeminists, who see a close connection between feminist and environmental concerns, see as a large part of the problem. When interactions and relationships between the separated elements of the apart-from-nature model were recognized, they were interpreted as purely external relationships that did not affect the natures of the individuals involved. The related individuals were not seen as working parts of a system. Instead, they were ranked or graded in a kind of hierarchy that did not acknowledge the interdependencies of the parts of what is actually a functioning system. One independent thing was thought to be above another independent thing. Hierarchical thinking ranked mind over body, culture over nature, human over the nonhuman, and male over female. The ranked pairs came to include the wealthy over the working class and white people over people of color. Karen J. Warren calls attention to the hierarchism of the old model of nature; she calls it "value-hierarchical thinking" or "'up-down' thinking," which places a higher value on what is "up." Warren sees this as associated with "value dualism," which sees pairs of concepts as disjoined and not complementary, so that one is valued higher than the other, usually mind, reason, and male rated above body, emotion, and female. Warren disagrees with feminists who see all hierarchical thinking as inherently wrong; she realizes the importance of classifying and ordering some things in order to make judgments in daily life. Hierarchical thinking becomes a problem when joined with the "logic of domination" that usually went with it. Not only were males assumed to be superior to females, but this superiority justified domination of females. The "logic of domination" included a value system that sanctioned oppression.[7]

An important part of the problem with hierarchical thinking, it seems to me, is that the rankings of one thing over another were based on gross generalizations. It was not judging that this man is stronger than this woman, but judging that men are stronger (in all important respects) than women. This unfounded generalization was then used to justify domination. Hierarchical thinking was not the making of sound judgments when sound judgments were called for; the attack on hierarchical thinking should not become an attack on careful judgment. The thinking we criticize was poor judgment put to bad use. The aspect of this that affected the natural environment was the ranking of mind above matter, of soul above body, of a distant heavenly future above the present state of the world. This sort of thinking put man above nature, but left woman close to nature, to be dominated along with the exploitation of nature.

Patriarchalism vs. Sexual Equality

The subordination of nature to culture and of woman to man, which Karen J. Warren sees in terms of hierarchical thinking, is expressed in the name by which I refer to the old model "*man* apart from nature." The word *man* was used deliberately because a striking feature of the old view of nature is its sexism. It was the male who was seen as apart from nature and superior to nature. The female was closely associated with nature, sharing the inferior status assigned to nature.

The patriarchalism of theism in nineteen centuries of our era, and lasting to a great extent until now, is significant. There are some indications that the role of women in early paganism was not as it has been under theism, even though few, if any, cultures seem to have been entirely free of patriarchalism. Studies of religion in the Near East during the third and second millennia B.C. show widespread worship of goddesses. These were not all earth and moon deities: a number of them were associated with the sun, stars, or planets. These were not all associated with fertility cults.[8]

What the popularity of female deities indicates about the cultures of antiquity is not clear. Some feminists see it as evidence of early matriarchal cultures. In her 1975 book *New Woman/New Earth: Sexist Idiologies and Human Liberation*, Rosemary Ruether refers to "the myth of primitive matriarchy," which she considers unhistorical because the evidence of matrilineal cultures comes from neolithic times and the evidence of mother-goddess figures comes from a much later period. She holds, however, that the social and economic status of women was higher in earlier cultures than it has been in more recent times. She identifies urbanization and industrialization as turning points in the degradation of women. Even in peasant life, with the economy built around the family, women continued to have important social and economic bargaining power in agriculture, the processing of raw materials, medicine, and marketing.

Some American Indian cultures were not as patriarchal as the dominant culture of Europe. *Black Elk Speaks* and other evidences of Indian culture show a culture in sharp contrast to the European culture in attitudes toward women.

The debate about the role of women in primitive cultures is interesting, but it should not divert our attention from the important matter. However it happened and whenever it happened, Western culture adopted a notion of male dominance, which was expressed in the philosophy of Aristotle, a patriarchal view that received clear support and incorporation into Christian theology in the writings of Thomas Aquinas, for example, in *Summa Theologica* Ia.92,I. Thomas taught that the female is by nature subordinate to the male.

The Church polity of Christianity and the family life of Christian communities were strongly patriarchal. Orthodox Judaism and Islam support an equally rigid theory and practice of patriarchy. One aspect of patriarchal theism is the view, perhaps expressed most clearly by Thomas Aquinas, that the normative human being is the male. The female was considered the incomplete male, a view expressed or implied in a number of ways in all three theistic cultures. It is hardly an accident that the generic use of "man" and masculine pronouns to refer to humans in general or to individuals of unspecified sex became the accepted usage in English and some other languages. The phrase "man apart from nature" clearly expresses two aspects of the dominant view, the attitude toward nature and the subservient role of women.

Rosemary Ruether points out the close connection between the attitude of domination of nature and the domination of women. She calls attention to the sexist language used by Francis Bacon in talking about the exploitation and control of nature. Women were identified with nature and shared in the enslavement of nature. The liberation of women and environmentalism both require a social revolution that overcomes the attitude of domination.

Other feminist writers have followed up on Ruether's identification of the oppression of women with the domination of nature and see that environmentalism and feminism are closely related. Some feminists are critical of deep ecologists because they do not seem to recognize the connection between the exploitation of nature and oppression of women. Michael E. Zimmerman says these critics accuse deep ecologists of decrying a "gender neutral 'anthropocentrism' as the root of the domination of nature, when in fact *andro*centrism is the real root."[9]

Karen J. Warren strongly affirms Rosemary Ruether's belief that feminism and environmentalism are closely related; an important aspect of this connection is the historical coupling of oppression of women and the oppression of nature. Warren advocates "ecological feminism" and holds that environmentalism is not complete without a feminist perspective. She sees feminism correcting male biases and shaping a contextualist, pluralistic, and inclusive ethics that will oppose any kind of social domination.[10]

Failure to treat both sexes as fully and equally human is now seen as unjust and as an inadequate basis for an enduring society. A truly holistic ethical system cannot ignore half the human race. Failure to develop a feminism that embodies environmentalism would be to ignore matters of great concern to women and those who care about women. Failure to incorporate feminism in environmentalism would be to rob ecological ethics of very important ethical insights. An ecological feminism is demanded by our holistic ethical approach.

Sources of the Apart-from-Nature Model

The question of blame for the beliefs and attitudes that have led to the environmental crisis has been a matter of serious dispute. Lynn White, Jr., started much of the discussion in 1967 with a paper in the journal *Science* on "The Historical Roots of Our Ecologic Crisis," in which he held that Judeo-Christian theology was largely responsible. This theology exalts the human and places nature in a subservient role, as expressed in the biblical story of creation, according to which Adam is the culmination of creation and the plants and animals were made for his use. This made possible the ensuing disregard for the natural environment and the ruthless technological development that has degraded the environment.

White's criticism of the Judeo-Christian tradition has been supported by other writers. John Passmore attributed to this theological position indifference to the suffering of animals and the failure to recognize a duty to animals. Aldo Leopold saw biblical theology as a reason why ethics, which grew to include broader and broader human rights, has been slow to include the natural environment in the compass of moral concern.[11]

While not all religious thinkers disagreed with White, many argued that White had not interpreted the Judeo-Christian view adequately. They argued that the natural environment was valued by Jews and Christians because God had created it and that the command to subdue the earth required stewardship of it. Laws in Exodus and Deuteronomy requiring care of the land and of trees and animals were cited, along with other passages of scripture that reflect an appreciation of the natural world. Certain aspects of the theology of creation were held to favor conservation.[12]

Two decades after Lynn White's paper, the discussion continues, with studies of the Bible and the theology of creation.[13] For the most part, attempts to answer White's charges have given way to explorations of the role of religion in environmental ethics and ways for the various religions to meet environmental challenges. Many theistic and nontheistic religionists seem ready to move beyond past expressions of religious faith to confront a critical issue of the present.[14]

It is hard to see the dominant theistic religious tradition as having had a benign effect on the environment. In fairness, however, it must be recognized that Judaism and Christianity did not invent all the negative views of the environment. Early paganism based on the cycle of the seasons was replaced by a sophisticated dualistic paganism that was a product of certain strains of Greek philosophy, of which Neoplatonism was the highest development. The Jewish and Christian theism that finally developed was a product of Hebrew ethical monism combined with Greek philosophy, mainly Neoplatonism, and similar philosophies that preceded it. The notion of the soul as distinct from nature must be credited to that element in Greek philosophy that was influenced by the Orphic religion. Early on in Greek thought, Orphism developed a dualism of mind and body, with the goal of human life being the ultimate liberation of the soul, after many reincarnations, from imprisonment in a body. This dualism was central to the religious beliefs of Pythagoreanism, through which it influenced Plato. It plays a strong role in the philosophy of Plato, who used the symbols and myths of Orphism to express some important aspects of his thought, including his philosophy of friendship and even aspects of his epistemology.

In the late Roman period, the same kind of dualism of mind and body that influenced Plato was the central doctrine of Neoplatonism, which held that the material world was a nothingness, an absence of being, that caused what appears in human life as evil. Neoplatonism influenced the theistic religions with its concept of the human as apart from the world, a soul temporarily imprisoned in a body and yearning to escape from the world. An extreme form of dualism, with deprecation of the world as evil, appeared in the various versions of Gnosticism. Even though Gnosticism was rejected as heretical by both Judaism and Christianity, neither religion seems to have escaped entirely the extremely negative view of the world and the body that was part of Gnostic teaching. In the theistic religions of the Roman era, Judaism and Christianity, and later in Islam, the human person was seen as a being that must live in an evil (or deficient) world but is not a part of this world. The soul (mind) of the human makes the human a creature apart from the world, with a supernatural destiny that is not dependent on the world.

In the modern era the religious dualism of mind and body was augmented for some people, replaced for others, by secular versions of a dualism of the mind and the body. Descartes's version was the starkest and, because of his status as a philosopher, the most influential. Descartes argued that the mind, which is thinking substance, is utterly different from the body, which is extended substance. All mental activity, all conscious thought, feeling, and willing are in the mind. The body is an unfeeling machine. Nonhuman animals, lacking minds, had no feelings at all, not even feelings of pain. The cold horse makes a noise, but this is no different from the creaking of the cold wagon. The human body is like a nonhuman animal; without a soul the human would not feel anything. This kind of dualism led to a radical separation of humans, who are nonmaterial minds, from the rest of nature, which is unconscious matter. Nature was thoroughly demystified by this approach, so that it could be examined and used freely.

When we have determined that certain theological and philosophical ideas seem to be unfavorable toward sound environmental practice, have we discovered the cause, or even part of the cause, of the abuse of the ecosystem? Ti-Fu Tuan cites examples of abuse of nature in ancient Greece and Rome and in China to support his contention that aesthetic and religious ideals seldom have a significant influence.[15]

Classical Marxists believe that the economic system is the engine that drives society and that aesthetic norms and philosophical ideas merely reflect the opinions of the dominant social class. Marxists would look for economic causes for the harm done to nature. Others see the cause in human faults or social movements. The environmental crisis has been blamed on various economic and social factors and on various human failings. In the early replies to Lynn White, capitalism, technology, and urbanization, along with greed and ignorance, were seen as the cause of environmental degradation. These factors are still blamed for irresponsible treatment of the ecosphere. The question of whether wrong thinking about the natural world had a deleterious effect on the ecosystem, and more importantly, is it still causing damage, is a difficult one. There were probably a number of causes of human degradation of the natural environment. We have good reason to question the notion that beliefs affect human behavior very much. It is not likely that many people deliberately gave expression to negative thoughts by harming the plants and animals around them. It seems that few people give much thought to what they do and, if questioned, point to acceptance of a behavior by other people. When serious questioning of one's behavior cannot be avoided, few are the people who are not accomplished rationalizers. Even if our unconscious minds do not cause most of our actions, we can usually avoid acknowledging our less creditable motives.

We should not assume, however, that only beliefs of which people are conscious can affect behavior. If beliefs about nature have any effect on the general run of people, it must be effects of which they are not aware. The lack of thought about reasons for behavior does not show that beliefs have no effect. The harm that people have done to the Earth was probably motivated by feelings and desires that were not traced to beliefs, but which were acceptable to the person because of beliefs. The effect of the beliefs probably came through people's individual world views and their lack of sensitivity to matters that affect the environment. If a person's world view has been influenced by negative feelings toward wild areas or by a belief that humans are not a part of nature, that person will not see the destruction of forests or the loss of wild animal habitats as a bad thing. Just as an uncultured person can enjoy a book burning, the person with little knowledge of the human place in nature can enjoy a dry-season forest fire in the Everglades. The person who has never thought about belonging to a community of many species of living things can readily toss aside a handful of trash or drop tangled fishline overboard. I do not believe the causes of harmful behavior can all be explained in terms of intellectual error, but I do not believe that beliefs about the natural world have no effect at all. We need to promote good thinking and good social conditions and good mental health. People can be alienated from the society in which they live, and this must cause much antisocial behavior; a person with no sense of belonging will hardly bother to keep the community free of trash. Other people, however, seem to be quite comfortable in the community, but they trash it with no thought of what they are doing to the community of life of which they are a part. These people simply do not seem to constitute the harming of the environment as something to be concerned about. They do not seem to see nature as something important to them.

The people who cannot see an unused field as a place for animals to live, the people who cannot appreciate a stand of trees for anything more than a woodlot, the people who cannot understand why some people protect wolves, and the people who see an oil spill as nasty tar that gets on people's feet at the beach are people who do not have a holistic world view that includes recognition of the complex web of life. They are probably not concerned about the way forest creatures live because they know very little about the lives and the needs of these animals. They think little of pollution of the ocean, because they do not easily visualize what is happening beneath the waves. It does matter what people think, especially those beliefs that float below the surface of consciousness, because what people have in their world views determines what they see, how they see, and how they feel about what they see. When the person whose world view pictures humans as apart from nature, with only a temporary need for the world, when that person looks at a patch of woods, what will be seen? Will it be wasteland, land being wasted? Will it be seen as something ugly, something dangerous? It is not likely to be seen as an asset to the community, a place of rare beauty, or a source of peace and a pleasant change from the regularity of city streets.

Why does it matter what people see when they look at a patch of woods? It matters because actions follow from the way this piece of woodland is seen. The person who sees it as ugly, wasteful, or dangerous will not be inclined to support measures to preserve it. Cleaning it up, clearing it, burning its wood, and building upon it will be the only uses seen as fitting.

We can acknowledge that a number of factors contribute to the behavior that is harmful to the natural environment. The economic and psychological causes and human failures such as greed and ignorance do not lessen the need for efforts to help people overcome the wrong beliefs, values, and attitudes that corrupted the old man-apart-from-nature model of nature. World views can be changed, and presenting a clear and attractive new vision of the human place in nature is an urgent task.

CHAPTER TWO

Holistic Philosophy and Ethics

A New Philosophy of Nature

A new philosophical framework that sees humanity as an inseparable part of nature, and not as a spiritual entity apart from nature or even as a privileged part of nature, has been presented in several forms. These several approaches have much in common, along with distinctive differences. I believe there is something to be learned from all of them, but my own approach will be a bit different from any of them. A very influential approach was called the "land ethic" by Aldo Leopold, who pioneered a new ethic that held that humans have a moral responsibility to preserve the diversity and beauty of the natural environment. Arne Naess promoted what he called "deep ecology." Several writers advocated a biocentric ethic that was upheld as respect for nature. I am defending a critical holistic environmentalism I like to call the "person-planetary perspective," using the image in the title of a book by Theodore Roszak.[1]

The person-planetary image of humans in nature does not play down the worth of individual people or the worth of human culture; it recognizes the value of persons as well as the value of the natural environment in which they live. It recognizes the differences between human culture and the lives of nonhuman animals. It can be a very personalistic philosophy at the same time that it is a planetary view that takes full account of the planet on which people depend. Even more than this, it recognizes that people are an integral part of nature. It is not simply that we are totally dependent on this planet; we are, but there is much more to our being part of the land community. We are not alien spirits trapped in nature; we humans are as much a part of nature as all the other animals and plants. The ways we relate to the natural environment are different, of course, from the way nonhumans live their lives. We can shape our environment to our liking far more than any other organism can alter the conditions under which it lives. Yet we are dependent on this planet Earth. The person-planetary perspective is a holistic view that sees humans in interrelationship with the rest of nature, while at the same time it sees the special needs and capacities as part of that nature. Some aspects of the person-planetary perspective will be developed more fully in the following chapters. It will be seen to be feministic, which I consider an important part of being humanistic. It will be contextualistic and pluralistic in its ethical theory, in ways that will be explained later. In this chapter I wish to focus on the more biological aspects of humans being a part of nature.

The interrelatedness of humans and the rest of nature is much more than a practical reality that places certain limitations on us. The way humans are in nature is a more significant matter than many people realize. The role of humans in nature is not just a matter of location in an environment; it is not just the fact of dependency on nature. The person-planetary perspective enables us to understand what we are; it shapes the definition of what it means to be a human person. Until we recognize that we are part of nature, we cannot know who we are. This point was made by Arne Naess in talking about a "relational, total-field image" of the human's place in nature.

I have a favorite illustration for talking about this. An old automobile tire has been irresponsibly thrown into the trout stream. It is there in the stream, but it does not have an integral relationship with the trout stream. We can say everything important that is to be said about the tire without mentioning that it is in the stream. It might as well be in a dump somewhere. But the trout that lives in the stream is different. It cannot be properly described without reference to its environment. One of the most important things about it is that it lives in water. We humans are in nature more like the way the trout is in the stream. The old apart-from-nature model, or even a somewhat more enlightened human-in-nature approach, saw humans as being in nature the way the tire is in the water. Certainly the water has some effect upon the tire, and the environment has some effect upon humans, but there is much more to it than that. Thinking of people being in nature the way the tire was in water is not an adequate understanding of what people are. It does not give us a complete definition of the human person. Of course, there is another side to this; the human affects the natural environment, probably much more than the tire or the fish affects the creek. It is obvious that the fish has a narrowly determined niche in the creek, while humans can range over Earth. Nevertheless, the human is dependent upon Earth, even if a human can go from one niche to another. Visions of humans escaping from dependence on this planet are still, and I think will remain, science fiction; in fact, they seem to rest far more on fiction than on science.

The theme of humanity being defined ecologically is stressed by a number of writers on environmental issues. In the introduction to *The Subversive Science: Essays Toward an Ecology of Man* (Boston: Houghton Mifflin Co., 1969), Paul Shepard expresses this ecological thinking about humans in several graphic metaphors. He says the human skin is "like a pond surface or a forest soil, not a shell so much as a delicate interpenetration" (p. 2). He says that "man did not arrive in the world as disembarking from a train in the city. He continues to arrive" (p. 4). He speaks of the world as "part of our own body," noting Alan Watts's use of the same figure of speech (p. 3) in *The Book of the Taboo Against Knowing Who You Are*.

The pioneer developer of a person-planetary perspective on the land and its inhabitants was the late Aldo Leopold, who, following a long career with the U. S. Forest Service, wrote his influential *A Sand County Almanac*. The concluding section of Leopold's book calls for a new ethic, which he called a "land ethic." Just as the scope of ethics has grown from a time when people could be considered property, the disposal of which involved no ethical considerations, to include within the human community all people, ethics must now enlarge the concept of community to "include soils, waters, plants, and animals, or collectively, the land" (p. 204). Leopold's book became the primary manifesto of the new environmental movement.

The heart of Leopold's new ethic is his understanding of the land. Leopold held that land is not just a commodity to be bought and sold; rather, land is the foundation of all life. He described the land as a pyramid containing all forms and levels of life and as a "fountain of energy flowing through a circuit of soils, plants, and animals" (p. 216). This concept of the land is not a radical innovation in biological science. This concept of the land has its roots in the concept of nature as a vast economy within which each species of organism has its work to do, a concept that goes back to Linnaeus (Carl von Linné, 1707–1778) and which was developed by Raymond Lindeman and Charles Elton. Each part of the pyramid of life, producers, consumers, and decomposers (nature's own recyclers), plays a necessary role in sustaining life on the planet.

The holistic notion of nature as one economy has not been accepted by all ecologists. Some have argued that the natural environment is more a matter of independent "patches" than of one fully integrated system. Chaos theory and hierarchy theory have demonstrated that nature is not as simple and well-organized as some ecologists had thought it was. Environmental philosophers have taken notice of this development in ecological theory and are concerned about its implications for environmental philosophy and ethics.[2] I agree with Callicott's position, which is that the newer approaches to ecology are important and must be recognized by philosophers who are to have an accurate concept of ecological science, but that new theories do not show that a holistic metaphysical approach is mistaken. I do not see biological studies that employ a nonholistic paradigm, such as population biology, as disqualifying a holistic paradigm for studies of the relationships between species in a common environment and for other studies of energy circuits and trophic webs. Developing a model of the human roles in nature seems to me to depend mostly on the matters studied by those ecologists who do employ a holistic paradigm. I do not think the ideas of Aldo Leopold are passé, but philosophers need to be careful in dealing with these matters. If we are going to claim that ecological studies influence our concept of the place of humans in nature, we need to acknowledge changes in the field of ecological science. This is one reason why I refer to my holism as a critical holism. I do not wish to be bound to an early and outmoded ecological science.

Leopold said that the new ethic, the land ethic, must change "the role of *Homo sapiens* from conqueror of the land community to plain member and citizen of it." As citizens of the land community, we must not cause harm to the land community. Our treatment of the natural environment must be judged by the effects of our actions upon the environment. Leopold said, "A thing is right when it tends to preserve the integrity, stability, and beauty of the biotic community. It is wrong when it tends otherwise."[3]

Arne Naess made significant contributions to the shaping of a new environmental ethic. Naess's summary of his lecture at the 3rd World Future Research Conference in Bucharest in September 1972 outlines some of the significant ideas in a holistic approach to humans and their environment. Naess rejects a "man-in-environment" image for a "relational, total-field image," which stresses the interdependence of individual members of the biosphere. He refers to individual organisms as "knots in the biospherical net or field of intrinsic relations." This suggestive image can be faulted, I believe, for suggesting a too-fixed system of relationships and for understating the flexibility and self-determination of humans, but the attempt to form a "total-field image" seems to me to be a proper goal.

The second aspect of Naess's deep ecology perspective is "biospherical egalitarianism—in principle." This calls for the "respect, or even veneration, for ways and forms of life" that the ecological fieldworker develops; all forms of life should have "the right to live and blossom." It is important to note that Naess called for egalitarianism "in principle." He recognized that there are times when humans must use other species. Naess gave no indication of seeking an absolute equality, or of thinking it was a possibility. I am more comfortable with respecting the various forms of life and acknowledging their right to "live and blossom" without the misleading talk of equality. In fact, I am not sure that some organisms that are very harmful to other organisms should be guaranteed survival.

Confusion over the call for equality between species can be avoided, I believe, with a careful examination of the concept of equality. Even equality between humans in a democratic society is not a simple notion. Descriptive equality, which is equality in terms of some property that can be had in varying degrees, as when two cabbages are equal in weight, cannot be significantly applied to humans. There simply does not seem to be any descriptive quality that can be had in varying degrees that all humans have to the same degree. Evaluative equality, equality of some kind of merit, as when two books have equal literary worth, is of limited application in reference to humans. Human equality is usually interpreted by philosophers as a policy of treating people equally by making an equal, but not identical, contribution to their lives. Equality can also be seen in terms of an equal consideration of interests. These matters are not easy to determine in human relationships; their meaning in relationships with nonhuman animals and with plants will certainly be more difficult to act upon. Certainly environmental philosophers are not saying that plants and nonhuman animals are descriptively equal to humans. What can they be saying? The call for equality may serve most often as a challenge to an unthinking policy of always sacrificing the welfare of other species to the desires of humans. It is a rejection of what is called human chauvinism.[4]

The third aspect of deep-ecological thinking is support for the principles of diversity and of symbiosis. Naess says "the so-called struggle of life . . . should be interpreted in the sense of ability to coexist and cooperate in complex relationships, rather than ability to kill, exploit, and suppress." He speaks of ecologically inspired attitudes that are social as well as environmental, which favor "diversity of human ways of life, of cultures, of occupations, of economies. They support the fight against economic and cultural, as much as military, invasion and domination, and they are opposed to the annihilation of seals and whales as much as to that of human tribes or cultures."

The fourth principle is an anticlass posture, in human social relations and ecologically. Naess opposes exploitation and suppression of any kind, whether between social groups, nations, or species.

The deep ecology perspective shares with shallow ecology the fight against pollution and resource depletion, but Naess insists on a deep ecology perspective on this effort, one that takes account of all the principles equally and does not stress resources and the fight against pollution to the neglect of other evils.

The remaining principles favor "complexity, not complication," so that a "multiplicity of more or less lawful, interacting factors may operate together to form a unity, a system" and "local autonomy and decentralization," which will make various forms of life more secure, will promote self-sufficiency and reduced energy consumption.[5]

A number of other philosophers have helped develop the concept of humans as a part of the natural world, the person-planetary view. Paul W. Taylor has defended "life-centered" ethics, which he contrasted with human-centered ethics.[6] Others have used such terms as *non-anthropocentric* to stress the holistic focus of the new approach and the recognition of the moral standing of the biosphere itself. Others have used the terms *earth-centered*, and *ecocentric* to capture the essence of a holistic world view that rejects the concept of nature as "meant" for humans, as a resource only for human benefit. The complete development of holistic ethics is ecocentric ethics, which recognizes that humans have duties even to the nonliving parts of the planet, which are essential to life on the planet.

George S. Sessions, William Devall, Richard and Val Routley (now Richard Sylvan and Val Plumwood), John Rodman, the poet Gary Snyder, Pete Gunter, and a few others were early supporters of new approaches to nature and to environmental ethics.[7] They rejected as an arrogant human chauvinism the attempt to master and use the natural world for human purposes alone. Treating the world as though its only purpose is to satisfy human desires was considered inappropriate, not only because it went hand in hand with human misuse of humans, but because it is a mistaken view of the relation between human and nonhuman reality.

Conceptual Models and Metaphors

A number of different metaphors have been used to explain holistic concepts of the place of humans in nature. These metaphors are used to make the concept more graphic, but we need to look at the images with some care. Even though they are intended, I believe, to illustrate similar, and in some cases compatible, perspectives on the relationship between humans and the land, these several metaphors stress different aspects of the relationship and have different perspectives on it. They are analogies, and they have the advantages and perils of other kinds of analogical thinking. We should not assume that they are varied ways of presenting exactly the same model. To some extent, each one of them supports a version of the concept of humans as a part of nature. These analogies add a great deal of richness to our thinking about our place in the natural order, but some are more appropriate than others. We should try to find the images that fit most closely a scientific view of the natural world, avoiding charming images that are, none the less, poor analogies.

One of the earliest metaphors was the concept of nature as a vast economy, a concept at which we have already looked. This image of the natural environment stresses the interactions of various species of organisms and their contributions to the food chains or trophic webs that maintain life. This metaphor is compatible with Aldo Leopold's descriptions of the land as circuits of energy.

The metaphor of nature as a community was also used by Aldo Leopold, who spoke of a land community, and this has been a popular image with environmental ethicists. The community image has some significant virtues. It suggests cooperation in the pursuit of common goals and the securing of common advantages within an ongoing entity that is worthy of our work and sacrifices. It suggests added significance to our lives because we are a part of something that will not end when we die.

Those who use the concept seem to assume more similarity between human communities and natural systems than there really is. Human communities are deliberate and conscious associations subject to forces quite different from those that affect species' associations in ecosystems. Part of the attractiveness of the community metaphor is that, when compared to other models, it seems supportive of individual rights and freedoms. As attractive as the image of community is, we need to recognize that it has limitations as an analogy. In the natural environment there are species of organisms, even in the animal kingdom, that give no indication of acting freely as members of a community. If we follow the analogy carefully, we would see a community of very unequal members. The community metaphor, used to make the place of humans in nature vivid and understandable, seems to work better when handled as an abstraction, even a "prettified" picture.

Closely related to the community model, but suggesting even closer connections between the parts of the ecosphere, is the kinship metaphor. Some writers have used this way of picturing the relationship between people and the plants and animals to indicate that the relationship is more significant than a merely causal connection between organisms. There are a number of things that support the kinship language. Kinship is suggested by the marked similarity of cells and proteins in the human and other species, by the organic unity of life in trophic webs within one shared biosphere, and by the origin of species in a common process of evolution. Certainly this metaphor has its appeal; it uses a relationship that most of us find very comfortable, one that reminds us of pleasant associations. When we compare it to what really exists in nature, however, we see that the family connection is quite large; not only does it include some relations who are quite distant, which is typical of large human families, but it includes some who are so unlike us in significant ways that the analogy seems quite strained. The suggestive and striking image of kinship has to be used as an abstraction, perhaps even more so than the community model.

Another popular metaphor has been the comparison of the biosphere to an organism. This vivid model stresses the interdependency of all life forms. Individuals and species are seen from the perspective of their roles as parts in a vast system; the whole is emphasized and the parts are seen as subordinated to the whole. This metaphor has both strengths and weaknesses. Attributing to ecosystems the degree of unity and organization found in organisms is questionable. An ecosystem lacks the genetic basis of unity that makes an organism what it is. What does this metaphor have to commend it? It recognizes the great differences between species, which weaken the community and kinship models; it treats these differences as valuable to the working of the biosphere. Species are different, but their importance is determined by the role they play in the total order of things. The organismic model does not support human chauvinism; similarity to humans is not valued as it would contribute nothing to the biosphere.

Lewis Thomas defends the most extreme version of organicism, with his biological model of the biosphere as a cell. This metaphor diminishes the role of the individual more than any other image does. Thomas looks at individuals and societies as organisms, then sees them as cells in the organism that is life on Earth.[8]

There are those who object to organicism as a way of talking about the ecosphere, for obvious reasons: it seems to be contrary to a humanistic perspective. Human individuals are treated as no more significant than any other part of the system of nature, with aspects of life that are very important to people being given little or no place in this picture of human and nonhuman life. Frederick Ferré has recently expressed the "second thoughts" that a number of people have about organicism, but he looks to the difficult task of developing a concept of organicism that will not be antithetical to personalism.[9]

Garrett Hardin used the metaphor of life on Earth as a lifeboat to give force to some ethical issues, especially the threat of overpopulation. Just as the occupants of a lifeboat, which is designed to take a fixed number of people to safety, would destroy themselves if they took aboard a large number of people who are in the water, Earth is not able to take care of an unlimited number of people. Hardin advocated withholding aid to nations not willing to limit population growth, holding that failure to discourage the increase in the number of people would lead to vastly greater suffering in the future when the industrial nations could no longer prevent the starvation of large populations.[10]

There was a great deal of argument about Hardin's proposals. He was criticized for comparing the ecosphere to a lifeboat. The lifeboat image, as striking as it is, is not a good way to picture the world. A lifeboat is designed for temporary use, to keep people from drowning, starvation, and exposure until they can be rescued. The limited water and food aboard are not analogous to what occurs on Earth, which can recycle its resources. The social organization aboard a lifeboat is not like the society of plants and animals in relation to the land. Hardin may well be correct about the limited carrying capacity of Earth, and that was the main point he was making. Those who refuse to recognize a limit to the number of people Earth can sustain overlook some significant factors that indicate the rightness of Hardin's concern. But an analogy that can help make a limited ethical point may not be adequate as a picture of Earth. An image more analogous to the biological processes in nature will serve us better than the limited lifeboat image.

Another striking metaphor is that of Earth as a spaceship. This is better than the lifeboat to represent Earth. It pictures a scene of purposeful activity more varied than waiting for rescue. It emphasizes cooperation and the need to use limited resources carefully. In spite of its charm, this dramatic analogy has some severe limitations as a way to understand the natural world. Spaceships are engineered environments, machines, unlike the evolving Earth. They are not really self-sustaining; even the most advanced ships imagined by science-fiction writers are manufactured on Earth or on some other planet. They need many kinds of provisions, unlike Earth, which needs only energy from the sun and can recycle all its other resources. Spaceships are usually pictured as capable of wearing out or running down after their useful life. Of course, Earth is not expected to last forever, but the expected life of Earth is vastly longer than that of spaceships, and its dependence on the sun is quite different from the dependence of a spaceship on the design, fabrication, and provisioning by rational creatures. One limitation of the spaceship metaphor is the simplicity of function and social life of the spaceship in comparison with those of Earth. No spaceship is envisioned with the biological and social complexity of the planet Earth. The spaceship analogy can help us see some things about life on Earth, but there is much it does not show.

What is the outcome of our brief survey of metaphors for talking about Earth? We have seen that each of them makes graphic and vivid some aspect of life, but each has its limitations. None of the analogies fit the biosphere well enough to be adopted as our way of understanding Earth. We will probably be unable to find a perfect metaphor. Even faulty metaphors have their uses in making specific points, but they must be used with care.

We need to remember that behind all of the representations of Earth is the ecosphere itself, which we need to understand in as scientific a way as possible. The basic way of talking about Earth must be through use of the best scientific models and theories available at any time. I believe that scientific systems theory offers us the most useful approach. Cybernetic theory, which was first used to learn about the human brain through the study of electronic calculating machines, is a study of the way systems actually operate, with attention to how parts of a system respond to events in the system. The responses of the system to these events is called feedback. A study that leads to discovery of the actual feedback mechanisms at work in the world can help us avoid serious mistakes. I also believe that biological theory, with the confines it now has, must be augmented with the tools of other disciplines that study human life. The social, psychological, and intellectual aspects of life need to be understood better than we now grasp them. If we are really holistic, we must recognize that they, too, are part of the natural world. Chapter Three deals with this more fully.

The New Ethics

The person-as-part-of-nature view leads to a new kind of ethics. For the most part, traditional ethics treated the human person as an entity apart from nature, a nonmaterial mind or a spiritual soul. When the human person is seen as a part of nature, different things become centrally important.

Within the last three decades, we have seen a growing number of philosophers, legal scholars, social scientists, and people in environmental fields present scholarly papers on environmental ethics and related areas. Philosophers have explored the philosophical and logical grounds of new approaches to the ethical issues raised by environmental matters. The first explorations of environmental ethics stayed for the most part within familiar ethical traditions as they approached questions about human behavior toward the natural world. William K. Frankena argued that "all consciously sentient beings" should be morally considerable, but he saw no need to go beyond familiar deontological and teleological ethical principles. William T. Blackstone blamed the crisis of the environment on wrong values and attitudes. He urged fundamental changes in attitudes, values, institutions, and lifestyle, but he held that no new ethical principles are needed. What is needed is "a more sophisticated application of the principle of utility" and more careful attention to "justice and rights." John Passmore thought that the familiar principles dealing with injury done to persons would justify actions to prevent harm being done to the ecosystem. He said that what is needed "is not so much a 'new ethic' as a more general adherence to a perfectly familiar ethic."[11]

These ethicists were seeing environmental ethics as an extension of traditional morality to issues raised by the growing awareness of the deterioration of the natural environment and the limited supply of natural resources. Their approach to environmental ethics is called extensionism. The early writings in environmental ethics dealt with such questions as justice in the distribution of resources among nations and economic groups and the conservation of natural resources. There was some refinement of the traditional principles when they were extended to include concerns about animals and future generations. The concept of a moral community and the nature of claims and obligations had to be clarified but, for the most part, there was no attention to new ethical principles and new sources of moral insight.

In 1975 Holmes Rolston, III published an important paper in which he discussed the possibility of a new kind of environmental ethics. He noted an important difference between relying only on conventional ethical principles in the treatment of environmental matters and some recent approaches that found the basis of ethical judgments in principles developed in ecological science, such as the interdependence of species, the value to an ecosystem of biotic diversity, and a balanced order in nature. Rolston uses a 1969 paper by Paul Sears as an example of the earlier approach. Sears holds that scientific knowledge is an "invaluable guide" that is "relevant to making value judgments," but he does not see a possibility of a discovery of values through scientific knowledge; he says, "values are the business of the humanities." Science can only show what is necessary for human survival; this gives us what Rolston calls a proximate moral obligation—do that which preserves humanity, "assuming that is our wish."[12]

Rolston cited a paper by Thomas B. Colwell, Jr., as leading toward derivation of ethics from ecological knowledge. Colwell treats the principle of the balance of nature as "an objective normative model which can be utilized as a ground of human value." The balance of nature shows us "a naturally defined boundary" with which our values must be consistent.[13]

It was in Aldo Leopold's influential essay, "The Land Ethic," that Rolston found the primary use of such ecological concepts as the land community in the development of ethical principles. Rolston explains the structure of Leopold's ethic as the ecological law (recycling preserves the ecosystem) indicating a proximate moral law (you ought to recycle). The more basic moral law (you ought to preserve the integrity of the ecosystem) is derived from the value of a balanced ecosystem. What Rolston sees in Leopold's ethics is a "feedback from ecological science which, prior to any effect on the proximate moral oughts, informs the antecedent ought. There is a valuational element intrinsically related to the concepts utilized in ecological description." Rolston recognized problems and dangers in a new ecologically grounded ethics, but this project had great appeal for him. He said, "it has my vote to be so if it can."[14]

Rolston became one of the main leaders in developing the new ecological ethics; his clarification of issues relating to values in nature and senses in which people can and ought to follow nature is published in numerous papers in *Environmental Ethics* and other journals. Most of the articles are collected in *Philosophy Gone Wild: Essays in Environmental Ethics* (Buffalo: Prometheus Press, 1986). *Environmental Ethics: Duties to and Values in the Natural World* (Philadelphia: Temple University Press, 1988) provides a summation of Rolston's work in environmental ethics.

The concept of grounding environmental ethics in ecological science had a few more early champions. The scientist R. W. Sperry believed that science could become a source of values and high-level ethical principles. He said that "the grand design of nature" is intrinsically good, and it is right to preserve it and wrong to destroy it. The noted landscape architect and environmentalist, Ian McHarg, held that the direction of biological evolution reveals intrinsic value, with evolutionary success to be seen in diversity, interdependency, stability, and negative entropy. The philosopher E. M. Adams held that the world has a "normative structure" that provides objective value.[15]

Traditional ethics has generally supported ethical claims by appeal to utilitarian principles (which hold that it the duty of each person to promote the greatest possible balance of pleasure over pain for all people involved in an action) and formalist ethical principles (such as the demand that moral principles be intended as universal laws). Many writers on environmental ethics, however, do not consider these principles sufficient for handling environmental issues and argue for additional principles.[16]

The holistic view of right and wrong treatment of the planet gave environmental ethics a new ultimate moral value, the dynamic stability of ecosystems and the ecosphere. This is the heart of Aldo Leopold's land ethic, calling for preservation of "the integrity, stability, and beauty of the biotic community." Use of new ethical principles derived from ecological science raises new questions of justification, which require new attention to the logic of ethical judgments. The old approaches, which were developed to account for ethical obligations to humans, will not be adequate to deal with the new developments in ethics. The new sources of moral value and obligation and their incorporation into a system of ethics must be explained. Ecological ethicists are not willing to treat the value of such things as the integrity of the biosphere as simply a new taste to be adopted by those who like it; it is a significant source of moral obligation. They see knowledge of the interdependence of species and the presence of potential threats to the biosphere, along with the value of the stability of the biosphere, as reasons to place moral constraints on human beings.

Ecological ethicists see the integrity of the land community as a basic aspect of moral obligation, and they need an approach to ethical theory that can justify acknowledging its significance. Extending the traditional ethical principles to cover the new environmental issues has been rejected as inadequate by a growing number of environmental ethicists. The principles exressed by Aldo Leopold and the concept of a holistic ethical perspective have led to a new set of moral values and a new way of doing ethics. I will offer new ways of approaching the logic of ethics in Chapters Five and Six.

The Practice of the New Ethics

What are the practical requirements of holistic ethics? What behavior does it call for? What does it forbid?

Early on, Leopold, McHarg, Naess, and others explained what some of the actions required by holistic ethics would be. Morally acceptable behavior, they said, fosters the values found in the natural world,[17] which manifests a dynamic homeostasis that is seen in diversity and interdependence of life forms. Any behavior that threatens that stability is wrong. Human activity must be incorporated into the homeostasis of the ecosystem and not alter the system in ways that would upset its stability or reduce its diversity. A practical example of this is Ian McHarg's landscaping, which sought to make human habitations fit into the natural environment without destroying or greatly altering the patterns of life already there. I see what McHarg was doing as helping humans be plain members and common citizens of the land community. We have already surveyed some social implications of deep ecology that Arne Naess outlined. These were guiding principles for duties to the biosphere and for social duties. Paul Shepard and Theodore Roszak described what they thought human communities would be like if people lived in ways that would not harm the natural environment. Shepard favored a way of life similar to that of our hunter-gatherer ancestors, while Roszak favored placing humans in compact cities, leaving wild vast areas in the interior of continents.[18] Economists and political scientists spelled out other practical aspects of ethics based on a holistic approach to the place of humans in nature. These considerations of economic issues face squarely the finiteness of natural resources and the limited carrying capacity of Earth.[19]

Making actual determinations of which behaviors are harmful to the natural environment and which are benign is the job of ecologists. This has been a controversial matter at times; there are disagreements among biologists about such concepts as homeostasis and the role of species diversity. Bryan G. Norton has done extensive research on these technical matters and their implications for ethical theory.[20]

There are several areas in which the difference between the application of more traditional ethical systems, such as utilitarianism, to environmental issues and this new ethical approach based on ecology can be seen most sharply. One is environmental action, which has no immediate or obvious benefit to human beings but is beneficial to the environment. Some of these actions, such as wilderness preservation and limitations on grazing, mining, and timbering, will usually be contrary to the economic interests of some human beings. Curtailing land development to preserve animal and plant species, especially when these are not of obvious value to humans or are not beautiful, brings the difference between two approaches to nature into sharp focus. The argument often made for preservation, that the plants preserved might have medicinal use, possibly even be the cure for dreadful diseases, is seen by some ecologists as arguing from the wrong principles. Even if the holistic environmentalists accept the line of argument reluctantly as the only way to win public support for preservation, they are uncomfortable with that argument. The holist would prefer to defend preservation as a way to maintain a natural balance of nature in at least some places.

The disagreement between conservationists, who want to avoid waste of natural resources, and preservationists, who want to preserve wild areas in as nearly natural a state as possible, is one of the oldest areas about which holists and traditional ethicists differ. This dispute goes back to before the clear identification of holistic philosophies. The best known antagonists in this struggle were Gifford Pinchot, who established the U. S. Forest Service, and John Muir, a nature writer. Pinchot supported conservation and stressed the wise exploitation of natural resources for human benefit. Muir favored preservation and wrote of the beauty and spiritual benefit of wilderness. During their lifetimes, Muir's efforts seemed visionary, but finally the effort to preserve natural systems is winning strong support from governments and private interests. Florida, for example, has begun to restore the channelized Kissimee River, to protect the Everglades from agricultural run-off, and to protect from development numerous areas throughout the state. Florida was not the first state to begin such efforts. National parks and wildlife refuges are administered with an eye toward preservation as well as recreational use. The Nature Conservancy is saving wild areas throughout the United States and in other countries, and other private organizations dedicated to preservation of the natural environment enjoy wide support. No longer is preservation an impossible dream.

The new environmental ethics rests upon the realization that we humans are a part of nature. It is not enough to ask how nature's bounty is to be divided between human beings. We must ask how much of nature we have a right to use. We must ask which parts of the good earth we must leave alone.

Attitudes toward management of land and of natural systems show how the new and old ethics differ. Traditional ethics would see issues of land management only in terms of human interests; management in itself would be no problem. Deep ecologists tend to reject human management of nature entirely. Many environmentalists hold that human management cannot achieve what natural systems, acting with their natural rhythms and with natural feedback, can do. This is the real meaning of one of Barry Commoner's laws of nature, "nature knows best."[21]

Some environmentalists take the approach of leaving everything to nature, rejecting all human attempts to manage natural areas. A more moderate approach is taken by those who believe that humans have caused damage that nature cannot correct as well alone as with some very careful human assistance. Most holists are uneasy about humans taking over from nature and favor an incremental approach, with small-scale projects and time given to wait for feedback to indicate whether the desired effect is being achieved. Large-scale projects done in a short period of time have done great harm. The unforseen effects of large projects, such as the Aswan dam, are cited to support the claim that humans seldom foresee the full consequences of their actions. One of my favorite bits of environmental wisdom is that we cannot do just one thing. The change we have in mind is just one of numerous things that will happen because of our activity. The unexpected consequences of our actions are dismissed as side effects, but they are side effects only in relation to our main goal. Actually there are no side effects: everything we cause is an effect of what we do. Unfortunately, we do not know ahead of time all that we will cause to happen, especially when our project is a large one; we do not act with a delicate feedback system such as provides direction to natural changes.

The treatment of non-human animals exposes the differences between traditional ethicists and deep ecologists. Animal rights advocates are concerned primarily with the treatment of pets and domestic animals. They favor treatment centers for sick or injured wild animals, and they usually oppose the importation of exotic animals. Use of animals in research, factory farming of animals for food, and blood sports are some of the targets of these defenders of animals. The reason for the efforts on behalf of animals is to prevent pain or to recognize animal rights. Peter Singer and Tom Regan have taken different approaches to justifying actions that they both support. Singer bases protection of animals on utilitarian grounds, to prevent infliction of pain on conscious beings. Regan argues that those animals that have enough consciousness to be "subjects of a life" have a right not to be reduced to objects of human convenience. Both Singer and Regan oppose human activities that cause animals pain and a loss of liberty.[22]

Environmental holists have a different set of priorities. The efforts of defenders of animal rights to protect animals from pain is not considered as important as protecting habitat for wild animals in their natural roles in a stable environment. Pets and domestic animals do not have a place in wild nature, so environmental holists would see concern for the pain of such animals as simply a matter of sympathy, not of environmental ethics, which is concerned with maintaining the integrity of the natural environment. Some holists have expressed opposition to wild animal hospitals, which are rarely able to return an injured animal to life in the wild. The holists prefer that this effort be put into protection of natural habitat for endangered species. The environmental holist might not share the horror of hunting that the defender of animal rights is likely to feel. Clearly Aldo Leopold did not; he was an avid hunter.[23]

The differences between ecological ethicists and defenders of animals may be overcome or at least lessened. J. Baird Callicott has argued that duty toward the ecosphere does not remove duties on other levels, such as duties to family, friends, the nation, etc. He uses Mary Midgley's concept of the mixed community of humans and domesticated animals to account for our duty to these animals. Duty to protect the wild animals is a part of duty to the ecosystem.[24] In a paper titled "Can Animal Rights Activists Be Environmentalists?",[25] given at a conference on environmental philosophy and environmental activism that Lester Embree and I sponsored in May 1993, Gary Varner examined the assumption that animal rights activists and environmentalists necessarily will disagree on the issue of therapeutic hunting of those species that regularly strain the carrying capacity of their ranges, damaging the environment and endangering future generations of their own species. Varner finds in the philosophical principles underlying the views of Peter Singer and Tom Regan grounds on which defenders of animals can support efforts to control the populations of some species, which would require therapeutic hunting until other means of controlling population can be developed.

Traditional utilitarian and formalist ethical approaches are seen by most ecological ethicists as inadequate to deal with land use, wilderness preservation, treatment of animals, and other environmental issues. Holistic views of right and wrong treatment of the ecosphere give environmental ethics a new, ultimate moral value, the stability of ecosystems and the biosphere. The application of the ethics that values the stability of the planet as a good in itself to issues of land use, treatment of animals, and health concerns as well as political and economic matters is an ongoing exploration, as old issues and newly arising issues are examined in the light of ecological knowledge. The practical application of holistic environmental ethics will probably exhibit more changes from time to time than has been the case with traditional ethics. This is what should be expected with an ethical approach that is probably more attuned to what is actually occurring in the world, and has more expertise at its disposal, than has been the experience of previous ethical systems.

Justifying the new ethic, the philosophical task of showing that ecological ethics is logically sound and meets the other requirements of an adequate ethical system, will be treated in later chapters (especially Chapters Five, Six, and Eleven).

CHAPTER THREE

Holism and Individuals

Concerns About Holistic Environmentalism

The environmental philosophy and approach to ethics that has been developing during the past two decades has viewed human beings as part of the system of nature. Aldo Leopold urged the development of an ethic in which the concept of community was extended to include the land and all its creatures, with the role of *Homo sapiens* that of a "plain member and common citizen" who respects all fellow members of the land community. Arne Naess, an early and influential developer of holistic environmentalism, listed among the principles of the new natural philosophy "biospherical egalitarianism, in principle." Biospherical egalitarianism meant for him "respect, or even veneration, for ways and forms of life." Paul W. Taylor, who calls his approach "life-centered ethics," holds that human beings have no moral superiority to other animals.[1] Both Naess and Taylor accept, in practice, treatment of nonhuman animals and plants that, in effect, denies full equality of humans and nonhumans, so that the notion of equality has to be carefully examined and its implications determined. I treat this matter briefly in Chapter Two.

A number of other philosophers have espoused a philosophy that plays down the significance of humans in favor of nonhuman animals, or of all living things, or of the natural environment as a whole. Various names, such as ecological holism, deep ecology, biocentrism, and ecocentrism, indicate different but somewhat similar approaches to environmental philosophy, at least similar in that they all seem to deny a special place for humans in the scheme of things. Holistic philosophies, which incorporate in various ways the concept of the moral equality of humans and other species, are growing in acceptance by environmental ethicists and scientists and others with an interest in the environment. Biologists, especially, are acutely aware of the dependence of human life on the natural environment, and many of them find in nature precious but threatened values. Because humans cannot live well or very long without a healthy ecosystem, treatment of the environment is clearly seen as a matter of moral significance; for some this is a moral priority. Many people feel a moral commitment to the natural environment that transcends responsibility to provide for human biological need.

This acceptance of holism is not universal, however; some moral philosophers have serious reservations about the consequences of practicing this new approach to ethics. They have brought some serious charges against holism. We might wonder why, in the face of what is happening to the environment and the disasters that might be facing the human race, anyone would make a negative assessment of an ethics that takes seriously the well-being of the biosphere. Can anything be more important than the system of nature on which all life depends?

To put this into perspective, let us remember that previous holistic philosophies have been strenuously opposed by humanistic moral philosophers, whose main interests were human welfare, including freedom and justice. The holisms that made the nation the source of value and obligation are well-known examples. When the rights, the consciences, and the welfare of individual citizens were sacrificed to the good of the state, humanistic ethicists opposed such systems. An environmental holism that seems to put protection of the planet ahead of all else may be even more threatening than a social holism. Rather than a government requiring individuals to make extreme sacrifices for the good of society, environmental holism might require the human race to sacrifice itself to the good of the natural environment. If all matters must be subsumed under the health of the environment, will justice, the development of human potential in learning and the arts, the political equality of all persons, and other matters of humanistic concern be sacrificed or made secondary to ecological concerns? The stability of the natural environment does require changes in the way humans live. Human life must not destroy certain natural balances within nature. It is unlikely that a responsible person would deny this, but how far will environmental holism impinge upon human life? Will the finest fruits of centuries of humanistic culture be lost?

Some ethicists consider holism a threat to the rights of individual human beings or to individual animals. William Aiken found the practical implications of holism "astounding, staggering," because of the loss of individual rights. Tom Regan claims that "what holism gives us is a fascist understanding of the environment." Marti Kheel called ecological holism "totalitarian."[2]

Eric Katz writes about "the substitution problem," which is a result of failure on the part of environmental holists to understand the difference between thinking of the natural environment as an organism and as a community. If nature is understood in terms of an organismic model, the individual loses significance; one individual can be substituted for another if it makes no difference to the whole environment. This might happen when one species of animals can fill a biological niche and replace another species. Katz is disturbed by the possibility of this kind of substitution occurring when a matter of moral significance is at issue. To Katz, it appears that individuals are being valued for their instrumental roles in nature rather than for their intrinsic worth.[3] If the possibility of the substitution of one nonhuman animal for another or one nonhuman species for another is disturbing, certainly the thought of substituting one human for another as pawns in a natural system is horrifying. It will not be morally acceptable to most people to take into account only the role of humans as members of an ecosystem.

A number of feminists think that holistic environmentalism has failed to realize the historical connection between abuse of the environment and the denial of women's rights (see Chapter Two). Some fear that holism will actually encourage the continuation of the subjugation of women. Sara Ann Ketchum brought to my attention the fear that holism might tend to deny freedom in reproduction, which would be a special concern to women, who might be forced to have children, because both the bearing of children and the caring for children fall most heavily on them.

Some nationalistic holisms have limited reproductive freedom in the past, and even now some totalitarian nations deny reproductive freedom. Even though there is little likelihood now that environmentalists would approve of requiring women to reproduce to expand the world human population, women are not free to use effective birth control in many countries. Recently in Romania, women were under pressure to have large families for the sake of the state, and a few countries are still subordinating the freedom of women to what is considered a need of the country. In the United States feminists feel threatened by both foreign and domestic governmental policies that restrict access to birth control and abortion information. The excessive growth of human population poses severe problems for both governments and for the environment. Efforts to curtail the growth of population might lead to denial of reproductive freedom. Feminists hold that reproductive freedom is important, and even a possible threat to it is disturbing.

The concern of supporters of animal rights, such as Tom Regan, is that individual animals will be sacrificed to the good of the whole natural system. Holistic environmentalists have shown more interest in preserving animals in their natural role in natural habitats than they have shown in individual animals, which are the primary concern of animal advocates. Animal rights advocates talk about the pain animals are made to suffer and their loss of freedom. Holists talk about niches in natural environments; they do not express much concern for pet animals and farm animals. The holists do not show the same interest in conscious animals that can feel pain as do the animal activists. For the holists, not only are the animals with consciousness important, but also the lowliest decomposers in a food chain because the natural system needs the activity of worms and insects, even of microorganisms. The species that are most threatened are of most concern. The reason for this concern is not pain, which is a necessary part of natural systems in which every species eats and is eaten; the concern is the preservation of a dynamic balance within the natural system. The attractive, conscious animals that tend to look a bit like people are not favored more than the unattractive and even the inconspicuous forms of life. As we saw in Chapter Two, environmental holists and defenders of animal rights have different approaches to such things as hunting and hospitals for wild animals.

The concerns of environmental holists and animal rights advocates seem to be so far apart that Eric Katz and Mark Sagoff do not believe that the programs of the two are compatible.[4]

These concerns about holism are too significant not to be taken seriously. When responsible thinkers charge that environmental holism is totalitarian and a threat to individual rights, we need to examine the charges. If they are well-founded, they show that a holistic approach to the environment would require a rejection of humanistic ethics, with its concern for individual worth and individual rights. Before we support a radical rejection of centuries of moral thought, if we are to support it at all, we need to be clear about what we are doing.

Are the Fears Well Founded?

These reservations about holistic environmentalism are serious. Are they well founded? Are such dangers real? Is holism a genuine threat to individual rights and recognition of the worth of persons? When we examine the statements of holistic writers, some of them might strike us as antithetical to humanistic ethics. They do not sound personalistic; maybe they even sound hostile to human beings. Paul W. Taylor argues against the notion that human beings are morally superior to other animals. He says that an ethics of respect for nature is "life-centered," not human centered. George S. Sessions and William Devall do not seem to recognize a special place for humans, whose interests should be given consideration "only as a part of the larger whole."[5]

This certainly seems to be saying that people who do not play a significant role in the larger whole have no moral claim. J. Baird Callicott stated very clearly that holistic ethics is not friendly to humans. He says that environmentalism is biocentric to the degree to which it is misanthropic.[6]

Should the statements of holistic environmentalists be taken literally, or is this to be dismissed as rhetoric? A careful reading of these philosophers shows that they do not actually advocate the misanthropic measures their holistic principles might seem to call for. We note that Arne Naess advocated biospherical equality "in principle." He explained that equality could not be supported except in principle "because any realistic praxis necessitates some killing, exploitation, and suppression." Devall and Sessions think that the use of nonhuman nature for "vital needs" is justified. Even though the human population is too large, they say that we may take a thousand years to achieve the necessary decrease in the number of humans. Paul W. Taylor thinks that there can be adequate moral reasons for killing wild plants and animals, and he also says that humans are not obligated to make great sacrifices for the sake of wild creatures.[7]

J. Baird Callicott shows no inclination to support a misanthropic ethic. In several papers he holds that a holistic approach does not require rejection of humanistic ethical concerns. He considers the suggestion that human beings be destroyed for the sake of the environment "morally skewed." All that is required is that humans accept some moral obligation toward the biotic community and be willing to make some sacrifices for its good. He does not, however, suggest that obligations to the human communities to which we belong should be neglected. Callicott describes the several levels at which we have moral responsibility as nested rings of involvement in human society and in the environment. Our families, our circle of friends, our vocational associations, our towns and states, our nation, and the world, are communities, like rings within one another. We have differing levels of obligation within the several communi-ties. The immediate family gives us special obligations that we do not have toward neighbors and friends. Different still are responsibilities to the community and nation. The duties to the wider communities do not override obligations to the more immediate groups. Obligations to the whole human community and obligation to the biosphere do not override duties closer to home.[8]

Are the moderate, practical proposals of the present-day supporters of environmental holism enough to calm the fears of humanistic moral philosophers? Does the lack of misanthropic programs show that there is no basis for concern about holistic ethics? I do not believe that this moderation can give the supporters of feminism, human rights, and civil liberties the assurance they need. The moderation of holistic philosophers in this generation could be an effect of personalistic ethics that they have not been able to throw off. Can we count on environmental ethicists in the future maintaining the difference between policies that are approved for nonhuman animals (such as culling of starving animals) but not acceptable in the treatment of humans? Another generation raised on holistic philosophy might not be this hesitant to employ stern measures to protect the environment, even measures that would eliminate many features of personalistic ethics. The logical possibility of a really misanthropic ethical program still remains. The fears of the humanist rest on the possibility that draconian, inhuman measures are the logical conclusion of holistic thought. Eric Katz acknowledges that the danger he sees is a theoretical one; he finds no morally offensive cases of substitution in practice. His concern is about what might be the logical result of holism.[9]

If it is simply the weight of tradition that is keeping holists from advocating practices that would take away civil liberties and human rights, we should worry about future generations of ethicists. They might respond to the more severe problems of environmental degradation, which they are likely to face, with practical measures we would find shocking. They might not be restrained by humanistic sentiments from taking holism to its logical conclusion. A time might come when humanistic concerns about liberty, equality, and justice for humans seems far less urgent than the need to maintain environmental stability. Will holistic philosophers be more likely to make that judgment than philosophers who have not committed themselves to holism?

If holism is to be defended against the charges that it is misanthropic and will lead to the loss of much that we hold dear, it must be shown that draconian and inhuman practical measures are not the logical conclusion of holistic ethics. This is what I hope to show.

Different Kinds of Holism

My defense of holism will require a clarification of environmental holism, which will show that there are several types of holism, and I will further try to show that the types that could lead to antihumanistic practices are not acceptable on logical grounds, not simply because they are morally offensive to humanistic thinkers. There are several ways in which the ethical aspects of holism can be presented, and holistic duties can be presented in moderate or in extreme versions. Most of the literature about holism has dealt with it as though it were a rather simple doctrine. The distinctions I am making have not been discussed except in the debate over community and organismic models of the biosphere, which was treated in Chapter Two. The distinctions I am making are relevant to that debate, but they approach holism from a different angle and are more inclusive.

First let us consider holism as a scientific doctrine, which I refer to as scientific holism. Scientific holism is based on those findings of ecology that show the interrelatedness of organisms in natural systems and the connections between those systems. An early statement of this notion was in Barry Commoner's laws of ecology, one of which was that everything is connected to everything else.[10] While not all ecologists think in holistic terms of ecosystems, the view that nature is a system of interdependent organisms has strong support. The concept of nature as a vast economy, with different organisms making their own contributions to the whole system of nature, doing their own work, has been a basic theme of biology since the time of Linnaeus. Living things are related by trophic webs or food chains, as described by Raymond Lindeman, Charles Elton, and others. Within these webs, organisms of various species play roles of production, consumption, and decomposition. There have been several interpretations of this theme and disagreement about a number of points.[11] The basic concept, however, has been an important aspect of ecological science, which has explored in detail the interdependence of living things on each other and on the nonliving parts of the biosphere. The concept of everything being connected to everything else is a popular idea with environmentalists. Much environmental education makes this a basic theme. The interrelatedness of organisms is often expressed in figures of speech using social analogies such as the family relationship or kinship or citizenship. The concept has been expressed in the poetry of Gary Snyder, Robinson Jeffers, and Francis Thompson. The suitability of figurative expressions of the interrelatedness of species is an interesting topic of discussion; the analogies fit in some respects and not in others. Perhaps it is to be expected that the figurative expressions and popular literature will be much simpler than the actual biological conditions. This is the strength of popular presentations, for some purposes, but it is not without its dangers. There is greater variation in the descriptions given by ecologists. There are numerous aspects of the relatedness of individual organisms and of species.

The relatedness of organisms and species in ecosystems, which reveals the dependency of each organism on other organisms and on the land, was a basic theme in Aldo Leopold's description of the land as a community and as a pyramid of life, which he also described as a "fountain of energy flowing through a circuit of soils, plants, and animals."[12]

The interrelatedness of species can be seen in their development in a shared process of evolution. This resulted in many characteristics that species share. The similarity of body parts and functions and the similarity in DNA can be seen as reflecting commonalities in origin.

It is important to see that not all versions of environmental holism that claim to be based on ecological science have the same relationship to scientific studies, even when we consider only those ecological researches that study ecosystems. Scientific holism can be a carefully stated doctrine that claims only what biological research has actually discovered about the relationships between specified plant and animal species and between them and their environments. It can give an understanding of specific ecosystems by showing how various parts of the system are related to other parts of the system. When Barry Commoner said "everything is connected to everything else," he was basing this law of ecology on the science of cybernetics, which studies the influence of one event upon others in the form of feedback. He said that "an ecosystem consists of multiple interconnected parts, which act on one another. . . . (S)tabilizing cybernetic relations are built into an ecological cycle."[13]

Out of this study of positively identified relationships, holism as a general notion is developed; it then becomes a conceptual model, a very general guiding principle that structures inquiry, determines what is significant, and gives a framework around which the science is systematically organized. As a conceptual model or scientific paradigm, holism can provide an overall way of seeing the world for environmental education. By pedagogical use of this model, various bits of information are made to fit together and biological knowledge becomes more striking and understandable.

The relatedness of organisms can be stated in an exaggerated way, as in the poem by Francis Thompson that says "thou canst not stir a flower / Without troubling of a star." When an environmental educator says "every living thing is dependent upon every other living thing," it should be understood that this ambiguous statement cannot be a claim that each thing is known to be dependent upon each and every other thing, as Thompson's star upon the plucked flower. The statement has been generalized beyond what research has actually demonstrated. Philosophers can also make exaggerated holistic statements; in fact Paul W. Taylor admits to having overstated the idea of dependency in his paper, "Ethics of Respect for Nature."[14]

Exaggeration is not the only fault we must be careful to avoid. The form a scientific holism takes must be attuned to the best understanding of ecology available at the time. We must not go ahead as though the theories of Frederic L. Clements and Eugene P. Odum were universally accepted ecological science. We must recognize the effect of chaos theory, patch dynamics, hierarchy theory, and other new approaches to ecology. We cannot escape the complexity and the developing nature of ecology. Environmental philosophers are concerned about the effect of such new concepts as hierarchy theories on ethical and metaphysical doctrines, as well they should be.[15] A philosophical definition of holism must be open to development as ecological science develops. It must avoid becoming rigid and dogmatic. I believe that hierarchy theory shows that there are several heuristic ways to look at natural systems. There may well be several ways conceptually to model the roles of humans in natural systems. I do not see this as indicating that holistic perspectives on humans and nature are to be avoided, but it does warn us against exaggerating our statements of the oneness of humans and nature, and it warns us against proclaiming a particular version of holism as *the* scientific model of humans in nature.

Scientific use of a holistic conceptual model is surely quite valuable in stimulating and giving direction to research in at least some areas. It seems to be helpful as used in environmental education of children and the general public. Those who use it, however, do well to remember that it is a conceptual model and not a report of actual research. When used carelessly, as exaggeration for rhetorical purposes, there is a danger that it will "backfire." Sophisticated people may cease accepting as factually sound the claims of environmental educators. Those who profit from the misuse of the environment can attack defenders of the environment as being unscientific.

Holism can also be stated in terms of what is valuable. I will call this axiological holism. This type of holism ascribes value on the basis of usefulness to the biosphere. An organism plays a role in the ecosystem, and that is what makes the organism valuable. Anything that in turn is useful to the useful organism is also valuable to the natural order. Axiological holism appears in a number of papers, such as J. Baird Callicott's 1979 paper on animal liberation.[16]

Axiological holism can be expressed in various degrees, from a modest claim that one source of value, among others, is the contribution made to the biosphere to an extreme claim that contribution to the biosphere is the only significant value. Between these two expressions of axiological holism is the strong, but more moderate, claim that contribution to the biosphere is the main basis for weighing moral obligations. This claim would imply that an important contribution to the ecosystem could be an overriding consideration in ethics. Written expressions of axiological holism have not been sufficiently explicit to determine exactly which version of holism was intended. There may be no defender of the most extreme version at this time. Statements that might be read as extreme axiological holism, when read out of context, do not seem to intend such an extreme claim when seen in relation to other statements of the writer. Some statements in literary works, especially poems and essays quoted with approval by some holistic philosophers, might imply an extreme concept of value, but they are usually not expressed in specific terms of value.

Environmental holism can be presented in terms of moral duty. I will call this deontic holism. This approach to holism claims that there is a moral duty to be a responsible member of the biospherical community. The moral standing of nonhuman entities also comes from membership in the biotic community. A modest approach to deontic holism would see duty to the ecosystem as one source of moral obligation, among others. Treating duty to the biotic community as primary and overriding would be a stronger kind of holism. To treat duty to the ecosphere as the only significant moral obligation would be the most extreme version. I am not sure any ethicist at this time accepts an extreme approach that would reject the concerns of humanistic ethics. Such strong defenders of holistic environmentalism as Devall, Taylor, and Sessions do not seem to adopt the severe positions that are theoretically possible; even though some of their statements could be read as expressions of an extreme approach, other statements clearly indicate a more moderate view. Callicott, with his recognition of different levels of moral obligation, clearly does not think that the land ethic calls for an extreme approach.

Holistic writers seem to accept axiological and deontic claims as mutually supportive insights. At this early stage in the development of environmental ethics, the use of both kinds of holism does not seem very systematic, but some work has been done in explicating the connection between axiological and deontic aspects of ethics.

Peter Miller has a theory of value as richness that he relates to ethical obligation, but he does not give much explanation of the way richness relates to obligation.[17]

J. Baird Callicott has written several papers on value theory. He is not very explicit about the way values are related to ethics, but it seems clear that he sees the connection in terms of his ethical theory, which builds on David Hume, Charles Darwin, and sociobiological theory. It is a moral-sense approach that stresses the value to humans of social sentiments. The benevolent and social sentiments can be directly related to the valuing of the good of other persons and the natural world.[18]

Holmes Rolston, III has written extensively on values in nature. His main concern is the nature of values, and he does an extensive taxonomy of value, including both levels of value and types of value. He deals with the effect of values on ethics. Rolston suggests a view on the way value judgments affect one's sense of moral obligation. He speaks of value in nature generating duty toward nature. He says that "we follow what we love, and the love of an intrinsic value is always a moral relationship."[19]

I think that my treatment of value and the basis of duty, based on the effect of a person's individual world view on ethical beliefs, along with the volitional element in the person's constitution of the world, helps explain the mechanism by which axiological and deontic claims are connected. I will treat this more directly in Chapters Five, Six, and Seven.

Antihumanistic Holism

The different kinds of holism I have mentioned are some possible versions of the doctrine. I am not claiming that each of the possible positions I have indicated has had an advocate. I am interested in them as views that have been supported or that might be defended in the future. They provide a sort of spectrum that will help make the whole picture clearer. My main concern is assessing the possible positions to determine which of the deontic and axiological approaches are sound and which are not.

An examination of the spectrum of holistic positions indicates that what the humanistic critic fears is the extreme versions of deontic and axiological holism. One approach to holism that threatens humanistic ethics is the extreme deontic claim that duty to the biosphere overrides all other duties. The claim that all duty can be reduced to responsibility to the planet is even more threatening because it denies much that we value in human life. The other threatening approach to holism is the extreme axiological approach, which denies any source of value except contribution to the stability of the ecosystem. This denies humanistic and personalistic values by reducing human beings to their ecological role. Is the critic correct in fearing that these threatening forms of holism are the logical outcome of holistic ethics?

I do not believe that the extreme versions are the logical outcome of environmental holism. My reason is a logical one. In the first place, the scientific evidence does not fully justify these extreme versions. They may seem at first to be based on ecological science, but they are not. I am not basing this on the impossibility of deducing moral obligations from factual knowledge; I support an approach to ethical theory that holds that a deductive justification of moral claims is not necessary. What is lacking in the extreme versions of holism is adequate justification for claiming a connection between the factual knowledge provided by ecology and the reductive and exclusivist claims made by the extreme versions of holism. The extreme claims lack the fittingness that is the necessary link between ecology and ethics, which I will explain in Chapter Six. The lack of fittingness is related to two logical flaws of the extreme versions: they are extremely reductionistic, in that they reduce the significance of human life to involvement in the biosphere, and abstract, in that they do not rest on specific concrete and observable relationships and interdependencies. As I will explain, this is not just a matter of distaste for abstraction and this type of reduction. These are logical flaws in the extreme forms of holism. There are flaws of several types. Abstraction is flawed because its claims are not supported by evidence; it is theoretical formulation that has run too far ahead of its factual base. Reduction is flawed because it simplifies things by ignoring matters that are evident and that have been held important by generations of responsible thinkers.

The extreme forms of holism are logically flawed by a lack of internal consistency; in their attempt to support holism, they are contrary to holism. It is not holism to ignore all values except the value of an organism to the ecosystem, as the extreme axiological version does. It is not holism to reduce all obligation to the one obligation to promote the stability of the natural environment, as the extreme deontic version does. Holism must be inclusive to be whole. Holism simply cannot be so reductionistic as to leave important matters out of consideration. Holism cannot be satisfied with a partial picture of humanity.

The Logical Conclusion

It is easy to see how a person who has read the writing of leading holistic ethicists might assume that an extreme approach to holism would be the correct position. There is implicit in environmental holism a turn away from a position that rests on a narrow concept of human interests. Callicott wrote of a connection between holism and misanthropy. Several representatives of the deep ecology perpective have cited literary works that express strong antipathy toward human beings. Callicott cites as an example of misanthropy a statement in *Desert Solitaire* by Edward Abbey, who said he would sooner shoot a man than a snake.[20] The poet Robinson Jeffers explains that he wrote the first part of *The Double Axe* "to present a certain philosophical attitude, which might well be called Inhumanism, a shifting of emphasis and significance from man to not-man."[21]

It is not new for writers to speak disparagingly of the human species. Even Leonardo da Vinci was aware of humanity's poor record in taking care of the earth. Speaking of the destruction caused by humans, Leonardo wrote,

> O Earth! What delays thee to open and hurl them headlong into the deep fissures of thy huge abysses and caverns, and no longer to display in the sight of heaven so savage and ruthless a monster?[22]

Strong statements of devotion to the environment and nonhuman animals is understandable when we see the harm people have done and continue to do to the system of nature. We have good reason to fear for the ability of the planet to support the forms of life we have come to love in the face of human destruction. Anger and dismay are understandable emotional responses to the situation. That is not the same, however, as an intellectually sound response. I do not believe that logical consistency will lead holists in the future to abandon the humanistic and personalistic aspects of ethics. Logical consistency will not force a rejection of concern for civil rights, respect for individual persons, and a recognition of the duties we have to our families, friends, and the several communities of which we are a part. Environmental ethicists in future generations might adopt extreme protective measures, but it will not be logical consistency with holism that brings this about. The severity of environmental problems might force the acceptance of draconian measures, but a properly understood holism will not be the basis of this. Even nonholistic moralists might turn to extreme measures in the face of great threats to human survival. Not holism, but fear and desperation, would account for these actions.

We need to see clearly why extreme types of holism are logically unacceptable. A careful look at the extreme forms of holism reveals that they are extremely reductionistic, reducing human life to the barest aspects of relationship to the natural environment. These extreme forms also tend to be so abstract that they do not speak of the actual entities that we experience in our lived worlds and that our sciences study. Such an extreme holism would substitute for the richness of our lives in the world a world of abstract relationships and incomplete images of persons, animals, and plants.

When environmental holism makes the system of nature and the individuals within it into abstractions, it has gone away from the ecological science it claims as its basis. It has lost sight of the actual individual plants and animals that compose the system. In place of the relationships between actual organisms and the land community in which they live, we now have an abstract notion of interdependence. Symbols have replaced substances, and the study of nature is in danger of losing the control of its experiential grounding. An adequate ethical system cannot be informed by such an abstract conception of nature, one in which the parts of the natural system become hazily perceived symbolic ciphers. The individual person whose interests we would consider sacrificing for the common good is without individuality or personal character. Under these conditions of thought, the substitution problem that worries Eric Katz would be almost inevitable.[23]

Extreme abstraction goes hand in hand with extreme reduction. When the only value accorded to the individual person is that which comes from serving a useful function in the ecosystem, the person has significance only as a part of a greater whole, and only a part of the individual person is treated as the whole person. The rest of the person has been disallowed; the person has been reduced. When the only significant behavior of the individual person is that which makes a contribution to the whole, the ecosystem, the same reduction takes place. This kind of abstraction and reduction reduces each individual person to those aspects of being human that we all have in common. When all persons are alike, what reason is there for not substituting one for another?

What is wrong with a reductionist approach? Why is it not just one alternative among others, the right approach for those who prefer it? I hold that extreme abstraction and reductionism are not only unacceptable to our present state of moral sensibility but also unjustifiable assumptions. They become meaningless by losing touch with our knowledge and experience. The reduction and abstractness are serious faults that prevent the extreme versions from being acceptable on logical grounds. It not a matter of one's preference. These extreme versions are too seriously flawed to be the basis of an adequate ethic.

This sort of reduction has the same faults we have seen in other reductionistic views, such as defining people in terms of sex, religion, or position in society. It is like the mistake of thinking of humans only in terms of rationality or in terms of the person's relation to a deity and ignoring all of the other aspects of what it is to be human. The abstract reductionism of extreme holism ignores many things that are morally significant, but that is only part of its failure. It ignores vast areas of human life and experience of the world. Its understanding of human beings is so incomplete that it cannot be correct. We are biological organisms who are part of an ecosystem, but to define us only in biological terms is to give an inadequate understanding of what we are. We are not fully defined unless social, cognitive, psychological, moral, and aesthetic aspects of our existence are taken fully into account. It is possible that all these aspects of human life have a biological basis, but this does not justify defining human beings in the most narrow biological concepts, those that can be applied to any other member of the biotic community. That people are biological beings must broaden biology, not narrow our concept of people.

Reductionistic views are not acceptable as alternative versions of environmental holism. They must be rejected. This is not a matter of preference. An adequate world view must be scientific and inclusive of our experience of our common world. The extreme versions of holism are so reductionistic that they are the antithesis of holism. They do not really consider the whole because they leave out elements of vital importance. They do not recognize actualities of human life. Any adequate holism must recognize all the important biological, individual, and social aspects of being human. Reductionistic holism is not scientific, complete, or factually adequate because it attempts to treat some things as being what they are not and fails to recognize them for what they are.

The logical inadequacy of extreme versions of holism is paralleled by another kind of inadequacy, moral inadequacy. An ethical approach that does not take into account all factors that might be morally relevant is inadequate. It is marred by the incompleteness that comes with reductionistic and overly abstract ethical systems. In order to deal with all that belongs in the determination of an ethical issue, moral thinking must be grounded in an adequate world view. (The nature and function of world views will be explained in Chapters Five and Six.) An adequate world view must incorporate an understanding of humanness in its fullness. Every aspect of what is involved in being human must be considered. All of the qualities of human life must be recognized and their moral significance weighed.

Human vulnerabilities as well as human capacities must be recognized. The human susceptibility to pain is one of those features of human life that has great moral relevance. It is ethically significant that human beings can feel many different types of pain, psychological as well as physical. Humans can suffer disappointment, grief, loneliness, a sense of rejection or of being scorned, a feeling of having been treated unjustly, and many kinds of frustration. We do not know with certainty what capacity other animals have for feeling these many types of pain, but to whatever degree these organisms have a capacity for pain, it is morally relevant in determining ethical treatment of animals. To dismiss this aspect of ethics as placing too much importance on self-consciousness, claiming that it is a kind of chauvinism based on a human characteristic, is a mistake. One need not, and I think should not, make self-consciousness the only matter of ethical significance; it is, however, something that must not be left out of ethical consideration. Utilitarians may be in error in making the experience of pleasure and pain the only guide to ethical decisions, but no sound ethical system can deny the moral relevance of pain.

It is ethically significant that humans are rational and self-conscious beings. Some ethicists seem to treat this as a matter of little or no importance, rejecting attention to human self-consciousness as a kind of "speciesism." This happens largely in reaction to an equally inadequate approach to ethics that gives this aspect of humanness too great a role, even seeing it as an absolute distinction between humans and all other forms of life and making it a requirement for being morally considerable. To hold that humans do not have duties to beings that do not share this human quality is a reductionism that ignores the moral significance of a number of morally relevant matters, such as the capacity to feel pain, being alive, and being a part of an ecosystem. It is not sound moral thinking, however, to substitute one reductionistic approach for another. Self-awareness is morally significant, but not the only matter of significance. Exactly how important it is in relation to other considerations needs to be studied carefully in each case.

As we criticize traditional ethical systems, we should remember that they were developed in response to certain common needs of human individuals and of human societies. This is one reason for our being critical of them, because our ancestors may have had a faulty understanding of some aspects of human life, but it is also a reason why we must be very careful not to discard them carelessly; they were responses to real concerns that might be important still. To the traditional concerns about human personal and social life, we must add some broader concerns as we seek to discover the morally right responses to our impact on the natural environment.

As we take account of our role in the system of nature, we do not need to ignore the concerns of humanistic ethics. It would be an intellectual and moral mistake to reduce our understanding of ourselves to that of organisms in an ecosystem. We do have a large impact on our environment, and this is a matter of moral concern, but we should not reduce our range of ethical concerns to environmental concerns. It would be a mistake to ignore concerns about the natural environment and a mistake to reduce all of morality to extreme holistic environmental ethics. We must acknowledge the many levels of moral obligation, from family responsibilities and obligations to close associates to duties to distant peoples, future generations, and the biosphere.

One of the strongest warnings of the dangers of holism came from J. Baird Callicott, who pointed out the difference between environmental ethics and a concern for animal rights. Later Callicott saw a way to bring together these two concerns. He gave an excellent explanation of ethics that includes several levels of responsibility. Using Mary Midgley's concept of the mixed community, in which domestic animals are members of the animal-human community,[24] he showed how we are morally responsible for fair and humane treatment of pets and domestic animals. At the same time we are responsible for preservation of the habitats of wild animals and for the proper treatment of these animals. Callicott shows how the combination of ecological ethics and the historical association of humans and certain animals provides moral obligations toward both domestic and wild animals.[25]

Trying to take into account everything that is morally relevant forces us to face complexity, conflicts, and confusions, but there is no justifiable alternative to hard study and the making of hard choices. We can expect our traditional ethical concerns to conflict with our responsibility toward the natural environment, but there is no morally acceptable way to escape difficult decisions. We cannot justify making our decisions easier by being reductionistic. We must bring together concerns about the environment, concerns about liberty and justice, and concerns about the quality of human life. In doing this we must maintain a rational consistency, recognizing that what applies to humans will apply to many other animals. It is worth noting that Arne Naess, in his influential 1973 paper summarizing his lecture at the 3rd World Future Research Conference in Bucharest the previous year, paid as much attention to humanistic aspects of ethics as he did to environmental matters. He spoke of the effects on humans of excessive crowding and of the need for accepting diversity in human ways of life, cultures, occupations, and economies. He spoke of the evils of military, cultural, and economic aggression and of other forms of exploitation. He spoke of good and bad conditions of human labor. He was concerned about government and advocated local autonomy and decentralization. Naess's article, one of the most important manifestoes of holistic environmentalism, thoroughly integrated humanistic concerns and concern for the environment.[26]

We would see quickly what is wrong with a description of the game of chess that described only the board on which the game is played and did not describe the differences of the chess pieces. It is inadequate to see chess pieces merely as things that fit on the chess board. We must recognize that some of them move one way, others in other ways, and some pieces have more power in the game than others. In environmental philosophy we must describe both the board and the individual pieces; unless we do this, we will have a distorted view of the biosphere.

A nonreductionistic and concrete holistic approach seeks to see every part of the biosphere clearly and completely. Ecological science and other natural and social sciences must be employed to provide as adequate a picture as possible of organisms, species, and the nonliving parts of the environment. A concept of the whole biosphere must incorporate knowledge of the individual members. This use of science should not be a return to outmoded concepts of science or to questionable metaphysics. If we talk about knowing things as they really are, we must keep in mind that we are talking loosely, in respect to both science and metaphysics. Knowing that all knowledge must be grasped within a conceptual model must keep us from thinking that we can claim to have knowledge that is not colored by our world views. The concept of "objective" knowledge of "brute facts" lost its credibility and usefulness long ago. We can, however, avoid excessive abstraction and reduction in the way we think about the members of the biosphere.

A nonreductive holism requires a responsibly scientific basis in order to have an ethical wholeness, one that does not set aside any part of a person's moral responsibility.

How to face the conflicts and disagreements that occur when we make ethical judgments from different perspectives and on the basis of different world views is the problem faced in Chapters Ten and Eleven.

Will future generations of moral philosophers abandon the humanistic ethics that have guided humanity for many centuries? They might do so. Predicting the thoughts and actions of future generations of moral philosophers is not something I am competent to do, nor is it the concern of this book. We do not know what sorts of situations they will face. We might have some idea of what the environmental problems will be, but we do not know with any certainty how severe some of the familiar problems will be or what new threats may arise. There is much we do not know about future social and political situations, and we do not know what ideologies will be shaping human thought when generations to come must make important decisions. It may be necessary for them to take draconian measures to secure the survival of life on Earth. The point I wish to make is that if moral thinkers adopt ethical principles we would now consider unnecessary and even misanthropic, it will not be the result of sound, nonreductionistic holism.

This does not mean that we can wash our hands of all concern about the future. Draconian measures might result from our failure to take moderate measures today, such as reasonable action to control human population. When facing a crisis that threatens the existence of humanity, moral thinkers might sacrifice many of the humanistic values that are important to us now. This could happen even if they do not adopt a holistic moral perspective; even with a philosophy of nature that does not acknowledge that humans are citizens and members of the biosphere, future generations could take strong measures to preserve the human race. My point is that these harsh actions would be their response to an environmental crisis; the abandonment of humanistic ethics would not follow an inexorable logic of holism.

I advocate a humanistic holism, on the grounds that it is logically sounder than abstract and reductionistic versions of holism and that it is morally superior in that it takes into account important matters that should be considered in making moral judgments. It will not do to reject this notion of humanistic holism as an oxymoron. As amusing as they might be as word play, oxymorons do not signal logical impossibility. On a serious level of thought, we can see that the concept of holism and the concept of humanism modify each other when used in one phrase, indicating what concept of humanism and what concept of holism are being used.

A humanistic holism, which avoids the reductionism that ignores important aspects of our lives and the abstractness that has lost touch with the actualities of people and the natural system in which they live, is the intellectually and morally responsible approach to our lives and to the world. *Person/Planet*, the title of a book by Theodore Roszak, gives us a clear word picture of the goal of our moral endeavor. A person-planetary approach will preserve the evolving ethical standards by which we have guided our lives over centuries. At the same time, we will be able to bring into our ethical perspective the findings of ecological science and recognize ourselves as inhabitors of the planet. We need a moral perspective that will enable us to preserve life on the planet and to produce an enduring human society. If we avoid abstractionism, which loses sight of the concrete relationships that make up our biological lives on Earth, and reductionism, which loses sight of what we are as persons, we can act responsibly toward the planet and toward persons.

CHAPTER FOUR

Anthropocentrism in Environmental Ethics

Opposition to Anthropocentrism

The advocacy of humanistic holism, with recognition of the moral importance of certain aspects of human consciousness, runs counter to a direction that has been taken by many holistic environmental philosophers. These holists have rejected what they call anthropocentrism in environmental ethics.

The roots of this objection to anthropocentrism can be seen in Aldo Leopold's *Sand County Almanac*, in which he described a split between people who think of the natural world as natural resources for the use of humans and holistic thinkers who think of the world as a land community. Leopold called upon humans to cease trying to conquer the land community and to become plain members and citizens of it. Later Arne Naess made a distinction between "shallow ecology," which was concerned with conserving the land and protecting it from pollution for the sake of humans, and "deep ecology," which recognized direct duties to the biosphere and advocated biospherical egalitarianism, in principle (*Inquiry* 16 [1973]: 95).

It has become fairly common for holists who recognize that humans have duties to the biosphere to use the terms *anthropocentrism* or *homocentrism* to refer to the view that all moral obligations are directed to humans. One of the earliest uses of *anthropocentrism* of which I am aware was by George S. Sessions in a 1974 article and in his privately distributed informal newsletter, *Ecophilosophy*, started in 1976. In the 1974 article Sessions wrote,

> as the search for causes of the environmental crisis has tended to move to the deeper level of examining Western society's most basic assumptions and attitudes towards nonhuman nature, an increasing number of scholars have concluded that the anthropocentric orientations of our Western religious and

> philosophical traditions have played a major role in the formation of ideologies which have resulted in the crisis.[1]

Richard and Val Routley (now Richard Sylvan and Val Plumwood) used the term *human chauvinism* for the strong form of the view that morality is "ultimately concerned entirely with humans, with non-humans given value and moral consideration only if it is in the interest of humans." They claimed that this strong version of human chauvinism leads to the belief that humans are entitled to manipulate the world and its contents in the pursuit of human interests without regard for other species. The Routleys rejected the view that ethical terms can logically be applied only to humans, and they denied the view that humans have qualities that merit special moral consideration.[2]

A strong distinction between an ethic that acknowledges only the moral standing of humans and an ethic that acknowledges moral obligations to the biosphere was an important part of the papers of Naess, the Routleys, Sessions, and others. The Routleys associated human chauvinism with the classical economic theory of value, which is based on the preferences of humans, in contrast to the new ethic, which holds that nature has intrinsic value. This identification of anthropocentrism with classical economic theory has continued to be part of the attack on anthropocentrism.

Environmental philosphers have tended to associate themselves with anthropocentricism or with a view that attributes value to the natural world and supports a philosophy called by various names, such as biocentrism, deep ecology, or simply nonanthropocentrism. The division between anthropocentrists and nonanthropocentrists left little room for philosophers who acknowledge that humans have moral obligations to the natural world and who do not accept the limited concept of value in classical economic theory but still hold to a human-centered ethics.

This separation of moral philosophers into anthropocentrists and nonanthropocentrists is apparent in papers in the journal *Environmental Ethics*. In the inaugural issue (Spring 1979), Charles Hartshorne argued for "The Rights of the Subhuman World." Later in this first volume, Richard A. Watson argued that reciprocity is a necessary feature of morality, restricting rights and duties to humans and a few other animals (Summer 1979). Roland C. Clement (Winter 1979) and Anthony J. Povilitis (Spring 1980) criticized Watson for being too anthropocentric.

Some papers in *Environmental Ethics* advocated utilitarianism and pragmatism as adequate foundations for environmental ethics. John N. Martin (Spring 1979) argued that in most cases utilitarianism is an adequate basis for preservation of natural areas. Eric Katz (Winter 1979) replied that utilitarian arguments not only do not justify preservation but may prove detrimental to preservation because utilitarianism is concerned only with the satisfaction of human needs, and these might not coincide with the needs of the natural environment. Katz later opposed Anthony Weston's claim (Winter 1985) that pragmatism is an adequate basis for environmental ethics. He held that pragmatism is too anthropocentric to support environmental ethics (Fall 1987).

Also in *Environmental Ethics*, Alan R. Drengson wrote "Shifting Paradigms: From the Technocratic to the Person-Planetary" (Fall 1980), holding that the technocratic paradigm is based on human interests, while the person-planetary paradigm transcends homocentrism. Paul W. Taylor wrote "The Ethics of Respect for Nature" (Fall 1981), contrasting a human-centered with a life-centered approach to the environment. In this very significant paper, Taylor held that humans have no moral superiority to other forms of life. Some writers held that they did not find understandable the ascription of value to the natural environment, while others claimed that a human-centered ethic was adequate to preserve the natural environment.

The conflict between anthropocentrists and nonanthropocentrists continues in later issues of *Environmental Ethics*. Warwick Fox answered an attack on deep ecology by ecofeminists who claimed that it should be opposing androcentrism rather than anthropocentrism. Fox argued that it is important for deep ecology to attack anthropocentrism (Spring 1989). Arne Naess (Summer 1990) and Val Plumwood (Summer 1991) continue to defend a nonanthropocentric approach against attacks by Peter Reed (Spring 1989) and Janna Thompson (Summer 1990).

The distinction between anthropocentric and nonanthropocentric ethics continues to divide environmental ethicists. The papers cited are just a few of those that defend anthropocentrism and those that attack it.

The Accomplishments of Deep Ecology

What has been accomplished by the distinction between anthropocentrism and deep ecology? Has it contributed to deeper understanding of the issues faced in environmental ethics? One thing it has done is focus attention on the question of the moral standing of nonhuman species and the biosphere itself. A number of environmental philosophers have found in nonanthropocentrism and the view that nature has intrinsic value a way of according moral considerability to nonpersons. Even if this approach should not turn out to be the best approach to environmental concerns, at least there has been a serious discussion of the issues. The deep ecologists' challenge to traditional ethical thought has given significance to issues that needed to be explored.

It does not hurt to have deeply entrenched ideas questioned. The belief that humans are radically unlike any other species had not been examined very often. David Hume's view that human and animal mentality are much alike did not seem to have much effect on people's attitudes. Knowledge of the evolution of species and knowledge of the dependence of humans on other species did little to shake the confident assumption that morality was a purely human business. It took the claims of deep ecologists to make us see beyond the comfortable confines of old ways of thinking.

Within the conflict between anthropocentrists and ecocentrists, some common ground was discovered. It was established that we humans have duties to preserve the natural environment. We have come to accept duties that were unknown just a few years ago. No longer do we hear many calls for humanity to rise above the natural environment. The notion of a new level of human civilization in which science will make dependence on the natural world a primitive stage of our development is no longer put forward as great wisdom. The only disagreement now is over the kind of environmental ethic we will have. There is no serious thought of a nonenvironmental ethic. The deep ecologists drew the line between an anthropocentric concern for the natural environment and an ecocentric concern for the environment, leaving no room for an ethical denial of concern for the natural environment.

I feel sure that deep ecologists are due most of the credit for moving moral philosophers to the present stage. Ecological science alone could not make the moral impression upon us that deep ecologists, working on a high moral plane, did. I do not wish anything I may say to detract from the credit they are due or belie my admiration for what they have done.

Reassessment of a Distinction

Now that we realize beyond all doubt that humans are a part of the natural world, as dependent as any living thing on the complex webs on which all life depends, we can consider a stage of thinking completed. It is time, I believe, to reexamine the distinction between anthropocentrism and holism as ethical approaches to environmental issues. Without questioning the value of making the distinction in the first place, we may now examine the possibility that it will not continue to serve us. The distinction may have been exaggerated, or it may have drawn the line at a place different from where we need to draw it.

For a number of years, Bryan G. Norton has held that the dispute over anthropocentrism is not as significant as many people thought. As early as 1984 he argued that it is a mistake to think that a nonanthropocentric concept of value is what makes environmental ethics distinct. What is distinctive is that environmental ethics cannot be made individualistic.[3]

Norton claims that there is a "crucial ambiguity" in the concept of anthropocentrism. The belief that humans are the only valuers does not entail that humans can value only human things (133). The "last human on earth" example has been used to show that the last human survivor, if an anthropocentrist, might use the last days of life to destroy the world. Norton does not think this example demonstrates that a person acting solely on the basis of human interest would destroy the natural environment. The notion of a human interest is ambiguous. It does not distinguish between felt preferences and considered preferences. Considered preferences must be in accord with a rationally adopted world view, which includes scientific knowledge and human ideals (133f). Norton distinguished between a "strongly anthropocentric" value theory, which considers only the satisfaction of felt preferences, and a "weakly anthropocentric" value theory, which includes considered preferences. Weak anthropocentrism can include considered preferences that will support the preservation of the natural world. It can include criticism of the values that lead to exploitation of the world (134f). In weak anthropocentrism a considered preference for the preservation of nature is not based on the belief that natural things have intrinsic value, which Norton thinks is a mistaken approach (136ff, 138, 140).

To explain the distinction between strong and weak anthropocentrism, Norton used the difference between fair distribution of a trust, which distributes to the heirs the resources of the trust, and management of a trust to preserve it into perpetuity (142f). Weak anthropocentrism cannot be based entirely on considerations of individual interest because, in addition to "distribution," weak anthropocentrism includes "allocation," which must acknowledge an obligation to "maintain a stable flow of resources into the indefinite future" for the preservation of human life (143f). The purpose behind allocation is the preservation of human consciousness (146). Norton refines his anthropocentric approach in more recent work (see the Bibliography for some of this work).

The nonanthropocentrist answer to Norton is that the anthropocentrist might, or might not, value the preservation of the natural environment. The weak anthropocentrist might have no reason to be interested in caring for natural things.[4] This response to Norton seems to assume that anthropocentrism will adopt the value theory of market place economics. This approach treats all felt preferences as equal. Formally, this approach to value theory is anthropocentric, but substantively it is egocentric. This value theory holds that value is a product of valuation, a view which Norton accepts, but he rejects the egocentrism of basing the concept of value only on felt preferences and the treating of all preferences as equal. We should not accept the assumption that a formally anthropocentric theory of value must be combined with an egocentric view of what people will value. In the early Routley paper, "Against the Inevitability of Human Chauvinism," human chauvinism was described as "a form of group selfishness, group egoism" (45). As strange as the term "group egoism" is, the charge needs to be examined and answered.

The term 'anthropocentrism' has often been associated with a formally anthropocentric and substantively egocentric theory of value, the value theory of market place economics. It has been assumed that the preferences of an anthropocentrist will lead to behaviors which put the natural environment at risk. Is this correct?

Let us look anew at the concept of anthropocentrism. The etymology of 'anthropocentrism' suggests a concern for humanity, not an individualistic, egocentric concern; the word *anthropos* refered to humanity, not to an individual male. More significant than the etymology, however, is that an appeal to anthropocentrism as the support for what is presented as an ethical position seems to be putting humanity at the center of concern. The people-before-alligators plea implies a concern for humans, no matter that the claim may be spurious as well as misguided. In both the formal and the substantive aspects of valuation, it should be humanity which is the point of focus. This would, of course, mean that intersubjectivity plays a role in valuation, as indeed it does. It is not the individual alone who is the *locus* of valuation, for few people cut themselves off from the beliefs, tastes, and judgment of other people in the process of valuing. Even simple felt preferences do not always follow purely individual acts of valuation. Carefully considered evaluations are clearly not products of single egos acting alone. Norton and perhaps most philosophers believe that values come from valuing and that humans may be the only valuers, but this need not make human evaluations shortsighted. It need not make them individualistic. Human valuers can take full account of their membership in the human community and participation as humans in the interconnected webs that support all life. We have strong grounds for expecting people to do this and for holding them blameworthy when they do not. To believe that humans do the valuing does not require us to accept the belief that valuing is always individualistic and unelightened. An anthropocentric ethic, one that is truly anthropocentric, will place the highest value upon humanity, which rules out purely self-regarding considerations. An intelligently anthropocentric ethics calls for using the resources of humanity in making moral decisions; such ethics will not be focused on the narrow concerns of the self, and it will not be limited to the moral and intellectual resources of the self. Unless it is so focused and directed, it will not be ethical, by any standards. With virtual unanimity moral philosophers have refused to recognize a life of only self-regarding considerations as a moral life. They have also been in virtual agreement with Norton's belief that an ethical life will be based on considered preferences, not on felt preferences.[5]

There might be some objection to calling a view such as Norton's anthropocentrism and the view that I am developing anthropocentric. Both views seem markedly different from the picture of anthropocentrism often presented in contrast to ecocentrism, but what reasonable grounds can there be for preserving *anthropocentric* for an egoistic and ill-founded moral position? Do we need to keep *anthropocentric* as a term of reproach?

There will be objections that people will take unfair advantage of the notion of anthropocentrism to cover their egoistic assault on the natural environment by claiming that they are putting humans first. They have already done this and will do it again. Greedy developers claim to put people first, but we know that they are putting dollars ahead of everything else. We should expose them because as moral philosophers we are trained to examine the justifications people give for their behavior. This is not to say that we should be moral police. We are not responsible for making everyone morally good, beyond giving guidance to those who wish to be guided. Ethics can help honest people understand what honesty is and what it requires, but it is not the job of ethics to keep people honest. To control antisocial behavior we need strong environmental control boards, honest newspapers, and alert environmental activists.

Is Anthropocentrism an Adequate Basis for Ethics?

Philosophers may have caused some confusion about anthropocentrism by using poor representatives as their examples. Those who attacked anthropocentrism and those on the other side who balked at talk of intrinsic value have sometimes put forward poor specimens of expression of the view opposed, at least in the less technical literature.

Those who rejected ecocentrism have pictured the typical ecocentrist as a person who preferred other animals to humans and looked forward to the demise of humanity. At best the ecocentrist was some kind of moony "treehugger." The deep-ecologists also used some poor examples, taking the worst offenders against the natural environment as examples of anthropocentric thinking. Were the despoilers of nature really champions of humanity? Were they *anthropo*centric? Regardless of what they might have claimed, were they not petroleum industry centrists, lumber industry centrists, developer centrists, or even egocentrists? Had we thought a little bit about the meaning of *anthropocentrist*, I do not think we could have graced with the name anthrocentrism behaviors and attitudes that do nothing to help the human race. What good does it do humanity for people to remain ignorant of the biological nature of human life? How do we help humanity when we disclaim or minimize the findings of ecological science? We were calling anthropocentric people who were helping destroy, and surely not always in ignorance, the very basis of the continuation of human life into the distant future. The trouble with labeling the despoilers as anthropocentrists is that we were giving them more than their due, and we were not able to recognize the real anthropocentrists.

Deep ecologists seem to have assumed that anthropocentrism must entail a rejection of obligation to the natural environment. No matter how sincerely the anthropocentrist affirmed concern for the biosphere, it was taken for granted that anthropocentrism could not provide the basis for a workable environmental ethic. The most serious mistake, I believe, was not taking it seriously enough to examine thoroughly the possibility that an anthropocentrist could be a holist. Somehow the claim to be both anthropocentric and holistic seemed contradictory, but we had made them contradictory by our own insufficiently examined definitions. We simply did not take seriously the ethics of environmental philosophers who approached environmental matters from an anthropocentric position and expressed their concern for future human generations and a world that would make those future generations possible. All we needed to know was that they were anthropocentrists, and that meant that their ethics could not work, no matter how sincere they might be. I realize that I am putting this in rather simple terms, painting with a wide brush and slinging tar all over things, but I would rather make the point now and qualify it later than let it slip by in polite language. There were a few careful examinations and analyses of different environmental philosophies, but for the most part, *anthropocentrism* was used without enough care. Its role seemed at times to be more polemical than descriptive. I believe it came to hinder more than facilitate careful thought, much less dialogue.

I believe it is time to examine anthropocentric approaches much more carefully than we have done. This will involve at least two types of analysis. Philosophical scrutiny of actual positions taken, their logical consistency and the clarity with which they can be presented, along with their practical implications, should be one part of this. The other part would be examination of actual environmental situations to determine whether an enlightened anthropocentrist will be able to support the kinds of programs that a nonanthropocentrist would support.

Philosophically there does not seem to be any reason why the person who is committed to humanity would not be an environmental holist. The findings of ecological science seem to be so clear about the relationship of humans to the rest of the natural world that a holistic position is difficult to avoid. An anthropocentrist might be ignorant, but there does not seem to be anything in anthropocentrism itself that would promote or justify such ignorance.

Of course, we must do our philosophical analysis carefully. A small philosophical difference can be consequential. The question of intrinsic value might reveal an important difference between anthropocentrists and deep ecologists. We should examine this issue, but it is not a simple matter of one group accepting the notion and the other rejecting it. Philosophers who use the concept of intrinsic value do not agree among themselves about the role of intrinsic value in making moral judgments. Eric Katz and Anthony Weston had an exchange in *Environmental Ethics* (Winter 1985, Fall 1987, and Fall 1988) that indicates some of the complexity of the issue. Weston holds that it is a mistake to base environmental ethics on intrinsic value. Katz does not affirm that intrinsic value is the best foundation for environmental ethics. He rejects the use made of inherent value by Paul W. Taylor and Tom Regan as being outside the mainstream of environmental ethics, which he finds in the views of Holmes Rolston, III and J. Baird Callicott. Intrinsic value does play a role in the ethics of Katz, in that it limits the scope of instrumentalism in environmental ethics. What we find, then, is a range of positions from the rejection of intrinsic value to the use of the concept as the foundation of environmental ethics, with more limited applications of the principle in between.

In all of this discussion, there is no clear indication that the concept of intrinsic value in nature would be incompatible with an anthropocentric ethic or that it is necessary to an ecocentric ethic, as the views of Bryan G. Norton and Eugene C. Hargrove, cited above, make clear. The whole matter of the relation of value to obligation is still being studied. An interesting aspect of this is the attitude toward nature of an anthropocentrist who realizes that humans are a part of nature. Would the value of the natural world for this person be any less effective ethically than the value the deep ecologist sees in nature? Will there be any practical difference because of the philosophical differences?

When it comes to actual differences in attitude and behavior between the enlightened anthropocentrist and the ecocentrist, what might we expect? Will the anthropocentrist necessarily be less holistic than the deep ecologist in actual cases? Will the anthropocentrist's position be too individualistic or self-serving? Will the anthropocentrist be unable to support preservation of species that might not be of direct value to humans? These are the sorts of weakness ecocentrists have suspected in anthropocentrists, but are these fears justified?[6]

The use of fanciful and science fiction examples, such as the "last person on earth" case may not help us much, as Brian Norton pointed out. There seems to be little, if any, reason why an enlightened anthropocentrist, even if the last person alive, would want to destroy the planet. The usual reply is that the anthropocentrist would not have a reason to leave the earth alone, but I do not think this is a significant answer. People do things because they have reasons for doing them; they do not act just because they have no reason not to act. In trying to find out the practical effect of anthropocentrism, we will learn more from the actual decisions people make in real situations, not from thought experiments with imagined situations no one has ever faced.

Let us see if there are programs to preserve the natural environment that are supported by ecocentrists and not by anthropocentrists. Of course, two people who support the same projects may do so for very different reasons, but one of the fears biocentrists and ecocentrists seem to have about anthropocentrists is about what they will do, not what they will think. I suspect that this search will not turn up many good projects that the enlightened anthropocentrists will not support. Rene Dubos and others have pointed out that preservation of wilderness offers a number of benefits to humanity, including spiritual benefits, that should not be overlooked. If an anthropocentrist does not support preservation of the rain forests, it comes from surprising ignorance, not from concern for humanity. There have been sufficient discoveries of unexpected values in plants of rare and threatened species to make an anthropocentrist work for preservation. The Pacific yew is just one of the latest of many discoveries. Concern for human health, delivery from unnecessary pain and early death, avoidance of conditions that make life difficult and unrewarding, and the loss of beauty and the inspiration found in wild places seem to be more than adequate to make the anthropocentrist a trusted ally in our environmental causes. I find it increasingly hard to believe that putting humanity first will make an intelligent and knowledgeable person less concerned about maintaining the sort of ecosystems on which human life must depend. Obviously the anthropocentrist will support efforts to avoid pollution, and the need to avoid overpopulation of any species is just as obvious to a person who knows what is happening, whether one is thinking of preserving the biosphere or preserving humanity. The fate of humanity cannot be separated from the preservation of the biosphere.

Any person who does not realize that humans are part of the system of nature and that humans depend on the complex webs of interdependent living things might act in ways that harm the natural world. Not realizing that humans are a part of nature is ignorance, not anthropocentrism. What we need to fear is ignorance. What we need to promote is a holistic understanding of what humans are as part of the system of nature.

If we cannot point out philosophical problems in anthropocentrism that are actually problems with the system of thought and not incidental effects of ignorance, and if we cannot find actual situations in which knowledgeable anthropocentrists do not support needed environmental programs, we should take seriously the possibility that a continuing opposition to anthropocentrism creates an unfortunate division among environmentalists. I expect that what we are going to find is that the important struggle is not between anthropocentrists and nonanthropocentrists but between those who are informed and morally responsible environmental holists and those who are ignorant or irresponsible.

What can we expect to gain from a more careful examination of anthropocentrism than we have given it? In the first place, we can overcome some real confusion. We can go beyond the polemical use of the term to a better understanding of the concept. Part of what we will accomplish in doing this is avoiding using the same label for people who give no evidence of a significant regard for humanity and people who support environmental protection out of a concern for humanity. If we stop using 'anthropocentrism' as a pejorative term for those with whom we have disagreements in value theory, metaphysics, or in some other area, we will focus more clearly on the actual areas of disagreement.

A second advantage of a careful examination of anthropocentrism is that we may be able to make environmental philosophy much more accessible to people who are not professional philosophers. Concern for humanity is easily understood, and many people who do not get caught up in debates about intrinsic value can identify with it. If anthropocentrism turns out to be inadequate as a basis for environmental ethics, we will not be able to use it, but if it has been scorned for inadequate reasons, such as being confused with other attitudes and weaknesses that are not actually a part of it, we would do well to build an environmentalist platform on it. We might find in an enlightened commitment to the human race a way to overcome some of our theoretical divisions and be able to expound a more understandable and convincing rationale for environmental action.

CHAPTER FIVE

Knowledge of the Good and the Bad

The Is/Ought *Impasse*

What is new and different about environmental ethics? We have already seen that it is new in that it focuses on values that were not emphasized, or in some cases even recognized, in traditional ethics. The value found in nature itself, in its stability and its beauty, in the harmonious interdependencies that make a wealth of life possible, is a new and central aspect of environmental ethics. Environmental ethics is new, we have seen, in the duties toward the biosphere, which are its central feature. We can see these as substantive aspects of environmental ethics.

In this chapter we will examine the new structural or formal features of environmental ethics. Environmental ethics does not merely present us with new insight into values and moral obligations. Philosophers in the field of environmental ethics are developing new ways of doing ethics as well as new logical structures for the explanation of moral duties and the justification of these duties to the natural world and its inhabitants.

The old ways of justifying moral judgments have been recast, and in part cast out and replaced, by environmental ethicists. There have been explorations of new ways of explaining the nature of ethics and the logic of ethical reasoning. Part of moral philosophy has, from the beginnings of ethical theory in Greek philosophy, done more than make moral judgments. It has paid attention to the meaning of the terms used in moral discourse. Plato was concerned with the meaning of such concepts as justice, goodness, and happiness, and he used his metaphysical doctrines to justify his interpretation of these concepts and the important place he gave them in his ethical doctrines.

Defining the meaning of terms remains an important part of philosophical ethics, but some modern philosophers expanded this concern to include defining the nature of moral judgment itself, not just the meaning of the terms. What are people doing when they make moral judgments? It had been assumed by most thinkers that people were reporting something they knew about an activity when they said it is right or wrong. In this century especially, this was called into question. Some philosophers supported a view called emotivism, which held that moral judgements were actually expressions, or even just ventilations, of positive or negative feelings toward some behavior. Others, called imperativists or prescriptivists, held that moral judgments are thinly veiled attemps to influence the behavior of other people. If these interpretations are correct, a moral judgment is not really a statement that can tell us something about the nature of a behavior. Stating a moral judgment is itself a behavior that reveals something about the one speaking.

Not all moral philosophers were willing to give up the idea that a moral judgment intends to say something about the behavior judged, and they have presented several theories supporting a cognitivist position, an explanation of moral judgment as expression of knowledge or belief about the behavior judged.

The approach to environmental ethics that I explain in this book is not an emotivist or imperativist approach; I try to show that judgments of the rightness or wrongness of behavior toward the environment are not merely expressions of approval or disapproval of the behavior in question. Of course, an expression of moral approval of the preservation of a wild river or disapproval of clear cutting a forest is an expression of how one feels about what is being done to the environment. I believe it is more, however, than an expression of feeling. This feeling of approval or disapproval is not a mere whim or fancy; it is grounded in one's knowledge and beliefs about the world. Also, even though most environmental activists would like to influence the behavior of other people toward the natural world, an expression of disapproval of urban sprawl is not just an effort to get the city council to refuse a zoning change request. It declares a belief that urban sprawl really is wrong. It gives voice to a belief that strict zoning to preserve wild areas is really the right thing to do. There is something about wrong actions, I believe, that makes them wrong. There is something about right actions that makes them right. The rightness or wrongness is often a complex and subtle matter, however, and no simple explanation of the logic of moral judgment is adequate.

The most difficult function of moral philosophy is the one we will need to attend to, the justification of moral judgments. Most moral philosophers have not been willing to treat moral judgment as a matter of mere taste or preference. They have looked for some reason to hold moral judgments, some basis on which some judgments can be shown to be correct and others to be mistaken. This has been a difficult and ongoing enterprise of moral philosophers. Environmental ethics seems to some philosophers to make this task of justification more difficult because of the connection between environmental ethics and ecological science. The relationship between factual knowledge and moral obligations is the point of difficulty. It seems to most people that moral beliefs should be based on factual knowledge, but there is a logical complication standing in the way of a simple grounding of morality on factual knowledge.

The complication, which makes the justification of moral judgments more difficult, was pointed out by the eighteenth century philosopher David Hume. Hume saw that it is not logically valid to deduce statements of moral obligation, 'ought' statements, from factual premises. In a valid deductive argument, nothing can be in the conclusion that is not entailed by the premises. One cannot argue that roses are red, and people pay a lot of money for roses, therefore hot soup helps cure a cold. There is nothing in the premises about either soups or colds and nothing about healing. It is just as invalid to base a moral obligation on purely factual premises. Even though we feel sure the killing of fish downstream has something to do with the moral wrongness of dumping industrial waste into the creek, we cannot deduce the moral wrong from the facts of the case alone. We cannot validly argue (1) the Allied Chemical and Tool Company is dumping acid into the creek, (2) this waste acid is killing fish, (3) therefore, dumping the acid is wrong. Are our hands tied by a logical rule so that we can make no judgment of the dumping? No, I do not think so, but we must draw the connection between fish kills and moral wrongness properly. Hume was right about the mistake of drawing moral deductions from factual premises.

The logical problem that Hume recognized is described as a dichotomy between factual knowledge and moral obligation and is called the 'is/ought' impasse. A related problem is a dichotomy between factual knowledge and judgments of value, the fact/value dichotomy. These are somewhat different, but closely related, logical problems. Both of these logical impasses seem at first sight to call into question the basing of ethical norms on ecological data. The issue, put simply, is that normative or valuational statements cannot be deduced from factual premises. 'Is' statements do not entail an 'ought' statement. Are we unable to use factual information in making moral judgments? Some philosophers have thought so, but they were thinking in too narrow terms. A problem does not need to be solved in the terms in which it is first presented. This chapter and the next are about using factual information in making moral judgments, but first let us see a bit more about the *is/ought* impasse and other aspects of moral thinking.

How and on what level of thinking about ethics does the impasse affect moral judgment? It does not often affect the level of moral thinking with which most people are familiar, the level of deciding how to act on the moral principles in which we already believe. This kind of thinking follows a well-established pattern that is not affected by the *is/ought* problem. The traditional form of moral reasoning used to decide what is the right thing to do includes two types of premises. First is a general premise that states a moral principle. Then there is a particular factual premise, which points out the fact or facts about the action in question, that shows that the moral principle is relevant to the action. Consider the following argument:

1. Destroying a species of food fish is morally wrong.
2. Use of gill nets is destroying several species of food fish.
3. Conclusion: fishing with gill nets is morally wrong.

The first premise states a general moral principle about the moral wrongness of destroying a species of food fish. It does not say why this is wrong, but it assumed that the principle can be defended. In this case the loss of a food source is harmful to humans and the loss of a species can be seen as environmental damage. The second premise states the facts about fishing with gill nets, that it will destroy several species of fish. It reveals what it is about the action that makes it subject to moral condemnation on the basis of the principle stated in premise one. The two premises together have justified the conclusion that fishing with gill nets is morally wrong. Without the general moral principle, the argument would not be valid.

To argue that draining a certain swamp threatens the stability of an ecosystem and causes the loss of an endangered species, and therefore is wrong, either is not a logical argument or fails to state an implied ethical principle that holds it is wrong to threaten the ecosystem and its endangerd species. Actually, we usually use moral principles that are so well known that we do not state them. If someone tells you that it was wrong to drain the swamp because it will destroy an endangered species, this person has left out the general premise, that it is morally wrong to destroy the species. This does not bother you or the other person; it has probably been assumed that you know and accept the general moral principle. Why then is there any problem? It is because the assumption, that you know and accept the general principle, might not be correct. Sometimes people do not know that a person has accepted a principle, but they hope that the principle will not be challenged if it is not "put on the table." In any case, incomplete arguments can lead to misunderstanding. Not stating the moral principle works without difficulty when all parties to the discussion share a moral outlook about the matter in question, but we need to realize what we are doing. Certainly moral philosophers will want to be clear about the assumption of agreement about the moral premise.

Usually, an argument that has a suppressed premise can be made into a valid, standard-form argument by stating the general principle. This is not to say, however, that the completed argument would be sound, which would require not only that the form of the argument be valid but also that both premises be true. A sound argument, which guarantees a true conclusion, must have a valid form and true premises. Much of the argument about environmental issues involves dispute about the truth of factual premises. Is it true that the waste discharge killed those fish found floating belly-up a few miles below the factory? Perhaps the cold snap killed them. It often takes a lot of research to establish the facts, and sometimes they remain in dispute.

How can we determine that the moral premises of moral arguments are true? It is here that the *is/ought* problem usually comes into play. We do not want to keep going in a regress, supporting our moral principles with other moral principles without any firm stopping place. It would certainly be helpful if we could deduce our basic moral principles from factual information, but the *is/ought* impasse stands in the way. If we cannot logically deduce our general principles from factual premises, what can we do?

This would seem to present environmental ethics with a critical problem. In many cases of ethical reasoning, there are a number of moral principles that have been accepted for centuries. Our general premises are there for the using. No one is likely to ask us to justify our objection to lying, stealing, and abusive treatment of people. When we base our moral judgment on the doing of what will increase the welfare of persons affected by an action or on the justice of an action, we are appealing to familiar concepts, and most people will see our principles as good reasons for an action. On the other hand, in environmental ethics many of the principles are new. Few people learned them "at grandmother's knee." Many people have never heard them explained.

As we have seen, traditional ethical principles have not been thought adequate to support environmental ethics, so new ethical principles have been used to support environmental ethics, such as Aldo Leopold's statement, "A thing is right when it tends to preserve the integrity, stability, and beauty of the biotic community. It is wrong when it tends otherwise."[1]

How is this new ethical principle justified? What keeps it from being no more than a bare opinion or a sentiment of a nature lover? It is claimed that this new ethical principle was derived from the findings of ecological science; Holmes Rolston, III describes such a principle as a "feedback from ecological science," which "informs" the basic or ultimate moral principles from which more specific principles are derived.[2]

The significant question concerns the way ecological science "informs" the basic moral principles of holistic ethics. Can there be a "feedback" from ecology that really provides justification for ethical principles without falling into the 'is/ought' problem? I believe that there is a way, perhaps more than one way, to make the move from factual knowledge to moral knowledge. J. Baird Callicott; Holmes Rolston, III; and others have tackled the issue of justification of moral principles. I will explain an approach I believe does the most adequate job of explaining the logic of justification, but I have great respect for the thinking of my colleagues, and I will not be surprised if a combination of approaches finally proves to be the most acceptable explanation.

Factual Knowledge and Moral Principles

Ecological ethics is the term I use for a system of ethics that not only deals with environmental issues but also relates its basic principles to findings of ecological science. Such a new approach to ethics must be explained and philosophically and logically justified. This chapter will atttempt to do both. This endeavor has occupied most of my philosophical research for about fifteen years. Many philosophers are now engaged with environmental ethics, a number of them with ecological ethics. Considerable analytic philosophical work has now been done to clarifify the terminology and the issues of ecological ethics, and this work continues. A review of articles in *Environmental Ethics* since its inception in 1979 will give an overview of work in this area. My published contribution to clarification of the theoretical issues is mainly in articles in *Environmental Ethics*, which are early treatments of themes developed in this book.[3]

How is ecological ethics possible in the face of the 'is/ought' impasse? This logical impasse has seemed to a number of moral philosophers to rule out any form of *ethical naturalism*, a term for any attempt to base ethics on factual description of the world. The logical impossibility of ethical naturalism is almost a dogma with many twentieth-century ethicists, especially since G. E. Moore used the term *naturalistic fallacy* to label certain types of ethical theories that defined moral terms, especially the word 'good,' in relation to naturalistic factors such as "being pleasurable" or "being the object of a person's desire." Kenneth R. Pahel questions the common assumption that there have been ethical naturalists who actually believed that ethical obligations could be deduced from factual premises or held that criteria of behavior could be "true by definition." He calls such an approach *analytic naturalism*, and says the belief that some philosophers have been so obtuse as to believe this approach possible is a myth. He explains that ethical naturalists have been doing something very different from this.[4] While I tend to agree with Pahel that it is hard to believe that ethical naturalists did not understand the difference between factual descriptions and ethical norms, the myth is widely held and ethical naturalism is a bugbear for many philosophers. A frequent reaction to fear of ethical naturalism has been to renounce any attempt to base moral principles on factual knowledge or even to relate them significantly. I definitely want to relate ecological knowledge and environmental ethics, but this requires great care in explaining the approach to be used.

A commonly accepted approach to the logic of ethics is based on concepts of justification that would make ecological ethics impossible. The approach assumes that deductive argument, with a normative major premise, is the paradigm case of moral reasoning and that the 'is/ought' impasse does not allow the basing of normative judgments on factual premises. This raises a sort of wall between knowledge and moral obligation. One response to this has been the adoption of emotivism or prescriptivism, because neither of these approaches tries to justify the basic moral principles that provide the general premises of moral argument. According to these theories, basic moral principles are simply adopted. Because ecological ethics holds that moral principles respecting treatment of the natural environment are not simply preferences or tastes but are somehow supported by knowledge about the natural environment, we must show that this widely accepted insistence on deductive argument creates an unnecessary chasm between factual knowledge and moral obligation. Because the duty to preserve a stable ecosystem is stated in ecological terms, and the value of a stable system is based on ecological principles, the new ethic requires a new understanding of the connection between knowledge and moral obligation. It is necessary to show that the notion of a gap between factual information and duty has been mistakenly interpreted.

We need to examine more carefully the *is/ought* impasse. The impasse is the result of a logical difficulty; the problem lies in the logical relationship between statements used in a deductive argument. The impasse need not apply directly to moral insights and moral values as such, unless they are founded on deductive argument. I will explain later that insight into what is morally good and bad is more direct and immediate than deduction. The *is/ought* problem shows up only when deductive argument is used in moral reasoning. The problem I see with some traditional theories is the importance given to deductive argument.

I believe the role of deduction in moral thinking has been exaggerated. Seldom do people arrive at their moral judgments by deductive arguments. The traditional pattern of moral argument is an ideal type that appears in ethics books more often than it is used in actual decisions about behavior. More importantly, general moral principles are not reached as the conclusions of deductive argument. What I want to show is that this does not make basic moral principles arbitrary and unfounded. If we are to understand moral judgments, we cannot limit our inquiry to formal logic. I believe we will understand more if we enter an area of reasoning broader than deductive logic, an interface between logic and psychology. This suggestion frightens some philosophers, but I believe we can investigate matters important to the understanding of human behavior without blurring the lines between philosophy and other disciplines, even while we are making use of those other disciplines. We need to proceed with caution, but if we treat psychology as a foreign territory with firmly fixed (perhaps fortified) borders with philosophy, we will be making our intellectual life a great deal more neatly partitioned than is human life itself. What interests me especially in this connection are psychological studies of moral judgment in which there is a fusion of fact and value, a fusion of *is* and *ought*. This fusion will be treated in the chapter on value, in which the concept of a fusion is attributed to Abraham Maslow and other researchers whom he cites.[5]

Of course, psychology and logic are not the same; logical problems cannot be solved on the basis of psychological knowledge. What psychological study can do is give us a clearer understanding of what actually happens when we make moral judgments. It can show us when we do and when we do not employ logical deduction, thereby giving us a better understanding of actual moral reasoning and the actual making of value judgments.

For some time, the realization that perception of fact and perception of value are not radically different has been fairly standard in value theory.[6] This gives indirect support to a theoretical approach that sees moral judgment as more immediate than an exercise in deduction. Several environmental philosophers have understood that ecological ethics does not attempt to deduce obligation from knowledge about the natural environment. Holmes Rolston, III saw that the relation between fact and value is not so much circular as immediate. Rolston said that in an ecological ethic "an *ought* is not so much derived from an *is* as discovered simultaneously with it," that the facts are not logically or chronologically prior to the value but rather "the values seem to be there as soon as the facts are fully in." In Aldo Leopold's approach the *ought* is discovered simultaneously with the *is*. Rolston recognizes that the logic of this is "evasive."[7]

One of the influential people in shaping the holistic approach to environmental ethics, Arne Naess, saw that the new ethics would not be based on a deductive model. He said, "the norms and tendencies of the Deep Ecology movement are not derived from ecology by logic or induction. Ecological knowledge and the life-style of the ecological field-worker have *suggested, inspired, and fortified* the perspective of the Deep Ecology movement."[8]

I believe that an explanation based on a fusion of fact and value in a person's constitution of the lived world, the world as it is experienced rather than understood theoretically, will make clearer the immediate, undeduced awareness of moral obligation. Some aspects of this will be explained in the chapter on value. First I want to state clearly that in looking to a nondeductive approach to the justification of basic moral principles, I am not turning to intuitionist or emotivist approaches but to interpretations of ethical decision that involve an integral relationship between ethical choices and factual information.[9] This is not an attempt to deny that the *is/ought* impasse holds with deductive arguments. It is clear that descriptive statements do not logically entail statements of obligation, and the *is/ought* impasse remains a logical barrier to the deduction of moral obligations from factual premises. Moral judgment, however, does not often seem to be the conclusion of a deduction, except in textbook examples. Appeal to information in the making of moral judgments is seldom an obvious case of falling into the logical error of trying to deduce an *ought* from factual premises; moral argument is frequently enthymatic, with only the premise that is considered specific to the case being expressed. If objection is raised to the argument, a typical response is that there is, of course, "a lot more to it." The defense of moral claims is a complex and many-sided activity. The role of deduction in it is very limited. The person who objects that there is "a lot more to it" probably realizes that the moral claim rests on moral values and insights that have not been stated. Part of what the sense of moral obligation rests on goes deep into a person's consciousness of the world and of human life.

There is some connection between obligation and factual knowledge, but philosophers have focused on a too-narrow area of human consciousness in trying to derive obligations from factual knowledge. Trying to derive obligation from knowledge through formal arguments was a serious mistake, both logically and as an interpretation of what actually happens when people realize that they are morally obligated to act in a certain way. As we will see in the chapter on values, value judgments are not derived from factual judgments. Awareness of value is as primitive as awareness of facts. In our original experiencing of some aspect of the world, valuational and cognitive and volitional elements are fused. Values and a sense of moral obligation have their roots in the same original comprehension of the world in which factual knowledge is grounded. Recognitions of value and obligation are abstracted from the lived world; that is, the world as we experience it and live in it, in contrast to theoretical notions about the world. The raw materials of fact, of value, and of volition are embedded in our original experiences, as fusions of fact and value, and we draw them out and separate them into different sorts of judgments.

I remember when as a young boy I first saw a large body of water. It was Mobile Bay, and to me it was awesomely large, with an expanse of deep, colored water with a surface of constantly changing texture. The water lapped upon the shore, turned back, and again ran up on the shore. The wind blowing over the water seemed different from the wind that blew over the farmland of central Alabama. There was much here that could be described, but this first experience of the bay was not just an inflooding of factual information. I liked what I saw and heard and felt upon my face. I was thrilled by it and deeply stirred. Immediately I knew something else; I wanted to be near the sea. My days as a "landlubber" were over. I knew what I would do when I could select my place to dwell. All mixed together in that first sight of Mobile Bay were the material for factual statement, judgments of value, and a desire to live by the sea. This is just one of many experiences I have had of places and events in which factual, evaluational, and volitional elements were thoroughly fused, with no one of them having any priority over the others.

Important experiences tend to show this fusion of the material for factual judgments, value judgments, and judgments of what we want. When we try to base one sort of judgment on the other, we discover that we cannot do it satisfactorily. When we have made a factual judgment or a value judgment out of the material first presented to us, we cannot use that judgment to explain all the other aspects of our experience. What was once united in our lived world has been changed into different kinds of judgment by our separation of various aspects of original experience. We have changed these aspects of the world in the process of extracting them and putting them into different categories. Consider the object of consciousness we call a factual judgment. It is a fraction of the lived world, drawn from the lived world and formed into a statement of fact. It is a reduction from the world we experienced originally. It can be true, but it cannot be the whole truth. Try to tell someone about a mountain. You say how high it is and try to describe the colors. Are you going to tell about the mists and the vulture circling its craggy top? Why do you feel that you have not conveyed very much of your experience? Perhaps you begin to understand why Cézanne painted Mont Saint-Victoire over and over again. If you are to share what you experienced, you must speak of awe and excitement. Are the awe and excitement deduced from the height and color of the mountain? Don't you speak of them because they are original parts of the experience? It would be incomplete without them. They are as much a part of the experience as the facts about height and color. In fact, they may be more basic to the experience and the part remembered the longest.

This examination of the fusion of factual, evaluational, and volitional elements in our experiences might help us to see that we should not expect to derive our values from facts, because factual judgments are based on only a part of any significant experience. It might help us also see one reason why it is hard to reason with people about their desires. Clearly facts, values, and intentions are not unrelated, but they are not related the way elements in logical reaoning are related in lineal fashion. If we were to try to justify our basic moral principles, such as the rightness of preserving the integrity, stability, and beauty of natural systems, with logical arguments, we would run into difficulty we would not be likely to overcome. If the premises of our arguments were factual statements, we would run into the *is/ought* impasse. If our premises were moral principles, we would still face the task of justifying the moral principles employed. Instead of attempting to justify moral principles through arguments, I will follow this pattern: (1) first recognize the fusion of factual and normative elements in our most primitive and least theoretical constitutions of the world, (2) then explain the role of personal world views in formation of a person's moral beliefs, and (3) move from the psychology to the evaluation of moral beliefs. This last step will involve the evaluation of the world views on which moral beliefs rest.

World Views in the Constitution of Experience

What, then, is the logic of moral reasoning? What I have found most helpful in understanding the connection between knowledge and moral obligation is the work on the philosophy of perception, done mostly in this century. I find that this discipline, along with some psychological studies, offers me the clearest insights into the way moral judgments are made in relation to other aspects of a person's life. I also find in the philosophy of perception suggestive ways of understanding how factual knowledge helps to ground moral principles. Fortunately it is possible to benefit from some of the basic insights of philosophy and psychology without becoming expert in the sometimes difficult professional literature. I will use some of the widely known concepts to explain how moral judgments are made and how they are related to life in general, including a person's knowledge and beliefs. Many readers will recognize these or similar concepts from college courses in perception.[10]

An important part of my understanding of the relationship between factual knowledge and moral obligation is the role of a person's world view in both the perception of facts and the perception of values. The term *world view* will be used even though there are difficulties and ambiguities in the term, which are sources of possible confusion; no other term seems to work as well. The term will be used to refer to the individual person's perspective on the world, not to any collective perspective, such as has been referred to by such terms as *medieval world view* or *Western world view*. Of course, the individual's world view is influenced by broader, more generally held, world views. The individual is influenced by the culture in which he or she lives. A world view, as I am using the term, is a changing complex of notions that includes beliefs about the nature of the world and about human beings, along with a system of values and the person's hopes and expectations. Some of these beliefs will be clear and can be articulated; others are largely unexamined presuppositions. The person's world view is affected by the language in which the person speaks and thinks, by various tastes, interests, and desires. This world view is important because it plays a crucial role in the person's perception of the world. It influences, through the way the world is perceived, the person's sense of moral obligation.

A basic discovery about perception is that human consciousness is active. The notion of consciousness as passively receiving impressions from the outside world has been rejected by most scholars in the field of perception. Consciousness is always directed toward an object. The object to which consciousness is attending may be a perceived object in the world or it may be an idea, a memory, or a problem with which the person is toying. Some of the objects that engage consciousness have long been recognized as activities of the mind. We have not always been aware that consciousness is active when people are conscious of such things as tables and chairs, traffic signs, and other people's bodies. Consciousness is active when people are conscious of odors, sounds, and the taste of things. Consciousness is active when people feel that things are smooth or rough or hot or cold. Consciousness is active when people have kinesthetic awareness of the position of their bodies and of tensions, pains, etc. in their bodies.

People have usually thought of the things known through the senses as objects in the world, whether they are located outside our bodies as are tables and chairs, or inside our bodies as are pains. Inside or outside, they are worldly objects because human bodies are things in the world. We may finally be overcoming the notion that consciousness is passive when we are aware of these worldly objects. We have long realized that consciousness is active when people turn their attention to such things as daydreams, memories, emotions, or intellectual problems. These latter objects of consciousness were not usually thought to be worldly objects, but separate from the world as mental activities. Now most of us realize that the separation of mental activities from the world is unjustified. In any case, consciousness is always attending to things in the world or to things we might think of as activities of an "inner life."

We do want to maintain a difference between imaginings and perceiving of real things, because at times people confuse the products of their minds with things in the world. The important point is that in all cases of awareness, consciousness is attending to something. We are never simply conscious without being conscious of something.

The activity of consciousness in addressing itself to objects shows that earlier views about perception were mistaken. These views, still accepted by many people as a kind of commonsense view, took the mind to be largely passive in perception. It was described as a passive blank slate written upon by experience. We now know that consciousness is active in perception, selecting out of many possible objects of attention the object we will see, hear, taste, smell, feel, remember, or think about. Much more than the position of our heads influences what we will look at. The study of the various factors that play a part in our selection of the objects to which we attend is fascinating. Our selection is influenced by our beliefs and interests, as well as by such things as movement and the amount of light reflected by an object. For our purpose, the role of knowledge and beliefs in our seeing and understanding the world will be the important concern.

The activity of consciousness is not limited to selecting an object to attend to. Consciousness places an identity on the object to which it attends, making it, for the observer, an object of this kind rather than some other kind. Consciousness perceives the natural characteristics of the object, such as color, size, shape, etc. Consciousness also perceives the object in terms of what we might call cultural characteristics, such as the name of the object, the familiar uses of it, the attitudes people have toward it, and the value placed upon it in our society. When I see a mink stole, my consciousness has made it a mink stole for me. The natural characteristics of being fuzzy, brown, and about so many feet long are part of what my consciousness sees in the object, but I also see it as expensive, ostentatious, and an offense to animal lovers. These latter characteristics are the cultural ones, and not all people would see them the same way. Other consciousnesses would see the stole as chic, stylish, and a sign of the social standing of the wearer.

The cultural aspects of something perceived play an important role in the formation of attitudes toward it. I used to delight in the shallow edges of St. Joseph's Bay at Port St. Joe, Florida. To me, the stained sand, the things that had washed up, the animal life, and the shells and other detritus were fascinating. To those who preferred a clean beach with white sand, this bay shore was ugly. What made the difference? In part it was the difference between a person who wanted a beach for sunning and a person who liked biological field trips. In addition, there was the difference between someone who could see signs of life where others could only see trash and dirtiness.

A technical term often used for this activity of placing an identity upon a perceived object, making it an object of one sort rather than another sort, is *constitution*. Constitution is a convenient word, so I will use it, but it is necessary to bear in mind that constitution is not creating something out of nothing. The term must not be thought to imply that there is a creation out of whole cloth, such that we can make just anything we want to make of a perception. Constitution is the way consciousness attends to an object that is present to us. To see some ordinary thing, such as an apple, involves this activity of consciousness. Constitution makes the object an apple for me, which is not the same thing as making apples the way God or Mother Nature makes them. Without my consciousness, however, there would not be the apple in my world of awareness. The apple sitting with other fruit in a dish on the table was constituted along with the other fruit, the dish, and the table. My consciousness did not passively record all of this. If I saw the apple as a basketball, I would be a very sick man, but I could have been in a hurry and seen the apple as an undistinguished part of a table decoration. In fact I could have ignored it altogether as I looked for the mail on the table. We often ignore such things when we have become very familiar with them. How long has it been since you looked at the pictures on your walls?

When I talk about my lived world, I refer to the world as my consciousness has constituted it. Your lived world has been constituted by your consciousness. The worlds we experience and live in are our lived worlds. Your lived world must be very much like mine in some respects and unlike it in others. You probably see as tables and chairs, as apples and watermelons, the same things that are tables and chairs, apples and watermelons in my lived world. Grits and mangoes may not taste the same to you as they do to me. You might not like them as much as I do, but there is a good chance we both like peaches very much. You might not see as beautiful all the people I see as beautiful, but we would probably agree on the beauty of some people. We would see some things the same way because we share many aspects of culture, and our experiences have probably been similar in some respects. We have learned many of the same things, but you know some things I do not know, and I know some things you do not know. You might know toadstools as members of certain species, while I see sea shells with Latin names.

When we speak of the "real world," the tangible, actual world we feel we can count on experiencing on a regular basis, we are talking about our lived worlds, for these are the worlds in which you and I live. We have these real worlds because of the activity of consciousness in constituting them. Without consciousness these real worlds would not exist for us, just as most aspects of them do not exist for earthworms, codfish, and pineapples.

Speaking of the role of consciousness in constituting our lived worlds makes some philosophers uneasy because they see this as making all reality subjective. This is not a correct interpretation of my analysis of experience of the world, however, because it is only part of what is being said. There are aspects of the constitution of lived worlds that are personal or individual, but the total constitution is not wholly an individual matter, except in those cases we readily recognize as delusion, gross confusion, or mental illness. Only those people who are unable to live in the world of common human experience, so that we consider them ill, somehow cut off from reality, live in a world that is largely an individual constitution. The objects we ordinarily constitute are not wholly independent of an extra-mental world. This extra-mental world has some part in the constitution of objects. The objects we ordinarily constitute are not whimsical imaginings, with no relationship to anything outside our consciousnesses.

Our understanding of the real world is shaped to a great extent by science. When we perceive things in the world, scientific concepts play a role in our constitution of them. To varying degrees, depending on the person's interest in and knowledge of science, highly theoretical notions affect the constitution of what is perceived. I do not often constitute things in terms of subatomic particle physics, but certain discussions can lead me to think of an object as a system of atoms. When I look at a mountain range, I might think about its origin in the movement of plates of earth. When I look at an insect, I might constitute it as a member of a species. You might be influenced by sciences that did not interest me as much as geology and biology. Science has been a significant part of Western culture, and it influences the way we constitute things. Its influence is more pervasive than, say, the influence of ballet, which might have something to do with the way some of us might interpret a shape.

With this talk about consciousness and the constitution of objects, I might give the impression that I am supporting a subjective idealism. Let me state clearly that I am not. Our constitution of objects in the world is influenced by culture, so that in a sense they are cultural objects and creations of human minds. But that is only a fraction of the picture. Constituted worldly objects are truly objects in the extra-mental world. We constitute the object in the first place because it is presented to us, and the identity we place upon it is not provided entirely by culture. Size, color, shape, odor, and other aspects of the thing are constituted as they are because of the object itself. If we are good observers, the object has a better chance of playing a significant role in our constitution of it. The more curious we are, and the more we delight in learning about our world, the more the world itself can give content to our lived worlds and to our world views.

Consciousness does play an important role in perception. Our only access to the world is through consciousness, but this is not the same as saying the world is nothing but consciousness. When we pay close attention to the ways the world is presented to us through consciousness, we see clear indication that there is an extra-mental world. We have ample reason to reject a subjective idealist concept of the world as ideas in our minds. Analysis of our experience of the world gives us strong reasons to believe the objects we experience, and undoubtedly many others we have not experienced, are parts of an extra-mental world.

One of the reasons talk of the role of consciousness in perception seems to make perception of the world subjective, in the sense that it is an activity of the human subject alone, is not taking into account intersubjectivity. When we speak of consciousness, we tend to think in too individualistic terms, thinking of consciousness as a unique and private thing. Actually the privacy of consciousness is very limited, and only a part of what goes on in the individual person's consciousness is significantly different from what goes on in many other people's consciousnesses. In studying perception we do not enthrone the individual consciousness and make it the arbiter of what is real. We learn about perception from seeing how consciousness operates generally and the objects it generally constitutes. We want to understand human consciousness, not just an individual consciousness. The isolated, individual consciousness is not the focus of study. What reveals most about perception is the common experience of the world, including the way people, not I alone, or you alone, constitute the world. What is of most interest are those constitutions of worldly objects that seem to be the same for all people, at least all members of a cultural group.

We have more intersubjectivity in regard to some things than we have with others. Not everyone knows the taste of mangoes, and some people like Brussels sprouts as much as others detest them. These differences we have learned to expect. The more important matter, however, is the large number of things we experiece in roughly the same way. We all find water wet and fire hot, and we see that floors are flat and tree limbs are round. There is an astounding array of things we constitute the same way. Part of this is a matter of cultural sharing. It does not surprise me that many people cannot tell a grunt from a tarpon. Are there things all humans see the same way? Do all Americans have some common perceptions? We can see there are degrees of intersubjective agreement.

Let us consider an example of different degrees of inter-subjectivity that I have used elsewhere.[11] Suppose that a copy of the Declaration of Independence fell from an airplane flying over the jungle many miles from the mouth of the Amazon River. An Indian, having had no experience of writing or paper, would find the thing puzzling, unlike any leaf or animal skin that person had ever seen. The Indian might be afraid of it or constitute it as a gift from the gods. Even with the wide cultural differences between this Indian and the founders of our nation, there is a degree of intersubjectivity, one probably shared with all human beings who have noticed the difference between flat and round things, rough and smooth things. We probably all see the sky above and the ground below our feet. We probably all find that we sink in water, but that it is pleasant to drink. We probably have a high degree of intersubjectivity because we are fellow human beings and we live on the same planet.

Had our paper fallen further down the river, an Indian with more experience of Western culture might constitute it as paper with an unknown writing. The degree of intersubjectivity is now greater. Had the paper been picked up by a Portuguese-speaking merchant in Manaus, it might be constituted as an official-looking document in a foreign language. An English engineer in Belém could read the words but might not know the history behind it. With the merchant and the engineer, both products of Western culture, we share an even greater degree of intersubjectivity.

There is always a limit to intersubjectivity, sometimes a troublesome failure in the people close to us to see things as we see them. This is more the case with cultural aspects of our seeing of the world. Even some Americans, who can know this is paper with words of historical significance on it, might constitute the copy of the Declaration of Independence as a page from the Constitution. Some Americans hold the Declaration in reverence and would be stirred anew by reading the ideals of a democratic nation, while other Americans would find the revolutionary sentiments distressing.

All the people who see the paper would constitute the thing from the sky as an object, fairly flat and smooth. From that agreement, the constitutions would diverge with differences in the people's knowledge of paper and writing and languages. Knowledge of history, or the lack of knowledge, would make a difference. In America some aspects of the constitution would vary with political views. There would be little disagreement in countries influenced by Western culture about the physical determinants of the object. About some of the physical determinants, there might be universal agreement.

This commonality in perception, called intersubjectivity, offers an important check against certain kinds of subjectivism. We want to avoid saying that whatever anyone perceives is true, which would be ridiculous. We also want to deny that the world is simply ideas in human minds, a view that has respectable antecedents but is not the outcome of careful study of perception. The simplest explanation of intersubjectivity is that we see many things, especially their physical determinants, the same way because we share a common world. We experience water as wet because it is wet.

If intersubjectivity does not convince the reader that subjective idealism does not follow from the study of perception, there are other indications in perception itself that there is an extra-mental world.

An important indication that the world is not ideas in people's minds is the spatial location of the objects of which we are conscious. The objects we see are always some distance away from us in a definite direction. As I sit in my most comfortable chair, my desk is to my left, not far away but not close enough for me to reach items on it. A bookcase is within reach on my right. In front of me is a sliding glass door, through which I can see "the jungle," as my backyard in affectionately known. Behind me are bookcases I cannot see without turning around. Things like desks and bookcases are always presented in this way, in a location apart from me. If I think about a bookcase, say one of those in my office at the university, the idea of this bookcase is not spatially located. I cannot exactly "place" the idea. No matter how I face, the idea is never behind me, nor for that matter is it quite in front of me. Someone might say that it is inside me, but it does not really seem to be inside any part of me. Ideas of things are so different from things actually perceived in regard to location that I cannot doubt that ideas are one sort of thing and worldly objects are another sort of thing.

We can observe another feature of perceived objects that is not a feature of thoughts. Objects that we see only present to us a part of themselves, that part that is turned toward us. When I lecture, my students can see only the front parts of me. I once said "excuse me" to a large photograph in a drugstore when I bumped into "her," and I felt rather foolish when I saw "Kodak" in large letters on the front of what had seemed to me a person. I should not have been embarrassed, however, because I only did what we do all the time. When only a front was presented to me, I constituted the object as having other sides. This usually works out well, as when my students take me for a three-dimensional object. There are very few objects that give us any trouble in this respect. We are presented a more or less familiar front and we constitute the whole object without giving it any thought. If we were to see a uniformly colored disk or globe set within a deep box, we would not know how to constitute it, but this is a very unusual situation. Usually we are able to constitute the objects we see as having parts that we have not seen, but when something fools us we are reminded that we actually do supply much that has not been presented to us.

Thoughts are different from perceived objects in this regard. We are presented all sides of things in thoughts. Ideas do not present one face and leave us to supply what was not presented. My idea of my briefcase presents back and front, inside and outside. I do not need to turn it around, open it, manipulate it or myself to think of any part of it.

We can actually examine some of the ways in which objects in the world play a role in our constitution of them. An interesting exploration of the role of objects in constitution is seen in the work of Rudolf Arnheim on art theory.[12] Arnheim explains the relationship between the lines and colors on a flat surface and the constitution of a picture of a three-dimensional object. The subject matter of the picture is not the concern here, because constitution of nonrepresentational and representational pictures shows the same influence of lines, masses, and colors. What is of interest is the way different plastic features on the canvas enable us to see pictures rather than just lines and colors. Arnheim shows how certain features of the artwork, what Eugene Kaelin calls surface counters, produce a unified object with a figure or figures seen against a background. Arnheim identifies certain "forces" in the artwork that tend to make part of the surface the figure, other parts the ground. Shapes with unbroken contour lines, textured surfaces, smaller surfaces, simple and symetrical forms, and the bottom part of pictures tend to be seen as the figure. The other parts of a picture tend to be seen as background. With these and other principles, and the interplay between them, we can account for certain objects seeming to be near to the viewer, others seeming farther away. The viewer's memories, beliefs, and interests affect the seeing of pictures, but these personal factors do not wholly account for the way the picture is seen. Commonly perceivable and readily described features of the artwork have an important influence on the way the picture is seen. Part of seeing an artwork is mental activity, but part of it is related to what is extra-mental.

The differences between worldy objects and thoughts are so great that we need have no doubt that there is a world outside our minds. The Everglades, about which I am much concerned, are actually in the world apart from my mind. I cannot simply think them preserved from the lack of seasonal water and the pollution from the agricultural areas and thereby save them. There is an important sense in which these trees and swamps and sawgrasses are more real than some other objects of consciousness, such as people's daydreams about winning the state lottery. Even important things, such as my idea about a birdfeeder that the squirrels cannot get into, are not like the key lime tree from which the squirrels jump into the present feeder. The lime tree seems to be there always, something I cannot doubt because of my assurance that it will be there to be seen, by me and by any other person in the yard. My idea of the squirrel-free birdfeeder is something no one else can see. If I talk about it, all I get is a skeptical smile. It still lacks detail in my own mind, such as the exact hardware needed to make it keep the seed dry but let me open it to put in new seed. This idea is not very real, joining the ranks of those noises I thought I heard, those things I thought I saw, and the anticipation of those dishes that did not taste as I expected them to. What is the common denominator of these less-real things? It is the lack of intersubjective verification. They seem to be too individually mine, too much a matter of my mind. Other things are accepted as more real than these things because most other people see them the way I see them, time after time after time. We base our sense of reality on a high level of intersubjectivity.

While intersubjectivity and certain features of the way worldly objects are constituted may give us sufficient reason to reject subjective idealism in favor of assurance that there is an extra-mental world, they do not go very far in helping us decide between incompatible constitutions of the world. We certainly do not want to accept all constitutions as correct because that would lead us into absurdity. Fortunately we do not have to accept all constitutions as equal. We can look critically at any claim about the world. There are several ways to evaluate claims critically, but the one I wish to explain is evaluation of the individual world view with which a particular constitution is compatible. If the world view is not credible, the constitution is subject to doubt.

A significant factor in the constitution of objects is the person's world view. The thesis a person places on an object, a situation, or an experience is compatible with this individual world view. Not only does a person's world view play a role in how the person interprets objects in the world, but a person's beliefs, interests, and expectations also have a great deal to do with the selection of objects to which consciousness is directed. There is an interesting correlation between what a person knows and what a person sees. The people who visit the wildlife preserve in my county do not all see the same things. One reason is that some know better than others what to look for. One person says, "There is nothing out there." Another spends hours finding exciting things. It is not surprising that people see what they are interested in. To understand better why people see what they see and see it the way they see it, we can look to the context of that person's world view.

Our constitutions, like our world views, include normative judgments and evaluations. If we see something as an opportunity for good, it is related to something considered good in our world views. If we see problems or dangers, it is related to something seen as dangerous or disagreeable in the world view. When we think of something as beautiful or ugly, or have other positive or negative feelings about it, a world view underlies the assessment. If a situation has ethical significance for us, or if it is seen as having no ethical importance, we have constituted it that way, and an individual world view has played a decisive role in this constitution.

The world view works most often without being noticed; we do not need to acknowledge it for it to have a significant part in our constitution of things. While it is working behind the scenes, we not only do not notice it, we would have a hard time saying exactly what it is. A world view is complex, so that our best efforts at understanding ourselves may not reveal all the sides of it. It is also dynamic, changing with our new knowledge and new experiences. Even though our knowledge is one aspect of our world views, we should not think of world views as storehouses of information. It is not the work of a world view simply to supply facts. The world view lays out the terms in which the world is experienced. When we reflect on our lives, we can come to understand part of the influence of our world view.

We do not need to accept our world views as incorrigible. With careful reflection we can come to see that some parts of our world views are out of date: we have held on to some attitudes or some misinformation we have outgrown. We have failed to register some new information to which we were exposed. We might be able to realize that some parts of our world view cannot be well accounted for. There might be notions we must have acquired in the past, before we were as scientific or as critical as we are now. World views, like old houses, need to be refurbished now and then, and some of them need a complete remodeling.

The Evaluation of World Views

It is important to realize that some world views are far less justified than others; there are a number of ways in which world views can be inadequate. World views are not innate; each one has its own history, and once developed it is not unchangeable. It is not unreasonable to challenge an individual world view.

On what grounds can a world view be criticized? Lack of basic consistency is one ground. A bit of inconsistency can be charming, but some world views incorporate seriously incompatible beliefs or values. Except for trivial matters, it is a serious failing when people can incorporate into a world view beliefs that cannot be logically held together. How can a person hold contradictory opinions? One reason this happens is that a belief can be vague, perhaps even a matter of words learned long ago and not really examined. The reasons people give for doing something are often amazing; they often provide very little justification for the action. A belief fondly held might have unrealized implications that are unacceptable. The consequences of acting on a belief can be unexpected. Some really determined people seem able to refuse to recognize an incompatibility that others find obvious.

Inconsistency is not the only fault to which world views are vulnerable. Other world views can be criticized because they are based on beliefs that are adopted in questionable ways. This happens when a person accepts uncritically the beliefs of parents, religious leaders, teachers, and friends. Even if these beliefs are true, the way they are adopted is not adequate, especially when a person feels excused from the necessity of study and research. A preference for ideas from questionable sources or an acceptance only of ideas that reinforce biases calls a world view into question. Obvious indications of the working of certain questionable psychological mechanisms should warn of possible inadequacy in a world view.

Lest I be accused of ad hominem argument, let me explain how I am using these evaluations of how a world view comes to be accepted. A distinction needs to be made between reasons for suspecting or supposing something to be the case and argument showing it to be the case. Of course, the most biased and untutored person might have happened upon a true belief, but it is unlikely, and a person's qualifications may be the only thing we have to go on in some circumstances. If an ignorant relative suggests one remedy and the physician suggests another, it would probably be wiser of me to do what the doctor ordered. Of course, my cousin could be right, but taking my relative's advice is a poor wager for me to take. When a person scorns the ideas of learned people or shows no respect for science, there is a practical sort of ad hominem that I recommend highly. There are reasons for not accepting a belief even when we do not have actual disproof of the belief.

Assessing world views requires good judgment. World views are usually not wholly correct or wholly mistaken. They are usually mixtures of sound learning and cultural nonsense. They are often incomplete, with fuzzy areas and blank areas in the picture of reality. Sometimes this is unavoidable, but at times it may result from an unwillingness to face unpleasant facts. There are people "who see the world through rose-colored glasses;" other people seem to be unable to see anything bright and promising. Beliefs should be accepted on the basis of better reasons than their hue or brightness.

A world view based on superstition cannot be trusted. The world view of a person who scorns science is not to be trusted. There is ample justification for criticizing the conduct of science; the history of science is a checkered one, and the scientific establishment does not always live up to its professed ideals. I am convinced, however, that the only substitute for poor science is better science. The opinion that denies weighty evidence and a clear conceptual scheme in favor of a murky idea without strong intersubjective verification is probably mistaken. The remote chance that a superstitious or unscientific person can have a correct world view is not worth fretting about. Science cannot do everything, and we should not expect it. An excessive adulation of science may be nearly as suspicious as the rejection of science. It is still the case, however, that scientific study is the only way we have to come to certain kinds of knowledge, including knowledge of the natural processes about which environmental ethics is concerned. To reject science in regard to those aspects of life and the world is to reject knowledge.

A biased and parochial world view has very poor credentials. The person who fears or scorns everything foreign, the racist, and the sexist have probably cut themselves off from many opportunities to learn.

A world view can be dishonest and self-serving. It is rather easy for people to make their views of the world fit the interests of their in-group or their personal desires. A more significant corruption of world view, however, happens without conscious awareness. Beware the views of the true believer, the company man, the champion of a social class, or the chauvinist. Irving Thalberg describes the mechanism, which he calls "visceral racism," whereby a person who is not consciously racist is affected in the way situations are seen and in what is expected of people.[13] For a long time we have known that self-interest can affect a person's judgment, but now we realize that it can affect a person's perception of persons, events, and situations.

Faulty world views result in faulty constitution of the world, much as faulty constitution of the world leads to inadequate world views. When we see that a moral position finds its support in a faulty world view, we should take note of it. We have more resources at our disposal for arguing about moral issues than we have been using. Does the criticism of world views give us assurance that we are right? Can we be certain? No, we cannot be certain, no more certain than we are in most areas of life. We can give reasons for our actions; we can argue for our positions. Having information, being honest with ourselves, being careful, consulting with those who might help us realize something we have not taken into consideration, and staying humble enough to do an about-face when we realize that our thinking has gone wrong are important parts of being ethical. We do not have to consider every opinion as good as any other; that would be an overreaction to the uncertainty we face in all aspects of life.

Have we shown how moral knowledge can be acquired? We have explained that moral beliefs are not derived from factual beliefs, so that moral judgments are not subject to the *is/ought* impasse. We have shown that basic moral beliefs are grounded in the same lived world from which factual beliefs are abstracted. We have indicated some of the ways we have for criticizing the world views that underlie our application of moral principles to our lives. Is this moral knowledge?

The answer to the question of whether we have moral knowledge will depend on what one thinks moral knowledge must be. If moral knowledge must be absolutely certain, then we do not have moral knowledge. If moral knowledge must be founded on indubitable premises used in a deductive demonstration, then we do not have moral knowledge. If moral knowledge is having the information and skills in judgment we need to make good moral decisions, then we do have moral knowledge.

There are some aspects of the ethics I am attempting to justify that a number of people are bothered about. I do not intend to avoid discussing these matters. In the chapters that follow, I will treat in some detail the matter of certainty and uncertainty in morality. I will talk about the contextualism that is a part of this approach to ethics, and I will explore the pluralism to which a number of contemporary philosophers have turned. I might not be able to assuage all the fears of those who would cling to a more traditional type of ethics, but I will not ignore them.

First, however, we need to see how one goes from moral knowledge to the recognition of specific moral duties. The basic values and moral beliefs about what is good and admirable, which we can abstract from our lived worlds and judge as to their adequacy, must be translated into moral duties. We must make the move from moral knowledge to moral obligation.

CHAPTER SIX

Deciding What We Should Do

From Moral Knowledge to Obligation

If the realization of moral obligation is not derived logically from the factual aspects of a situation, what is the connection between factual information and moral value? We have seen that fact and value are fused in the original experiences of a person's lived world, but this does not explain why we think we should do one thing and not do another. The fusion of fact and value does not explain moral obligations. When the world view with which a moral value is compatible is criticized, we are able to assess the moral value. This shows us what is good and bad, but that is not the same thing as knowing what is morally right and what is morally wrong, because the good and the right are not always the same. In explaining how we can gain moral knowledge, we have come only part way in our attempt to justify moral judgments.

We made an important move in taking ethics out of the *is*-implies-*ought* pattern of explanation and placing it in an *is*-with-*ought* perception of the world. What we accomplished at that point was adopting a psychological explanation that is closer to what actually happens when people come to think of something as good or bad. We had to go on from that point and find some way to judge whether these impressions of goodness and badness were correct or mistaken. One part of doing this is to evaluate the world view within which the ethical values appear to be appropriate. If the world view can be seen to be consistent, supported by a high degree of intersubjective verification, and arrived at in keeping with scientific methods and in a scientific attitude, we can consider it reliable. We are justified in trusting the awareness of value that is fitting in that world view. The fusion of fact and value that we discover has led us to adequate knowledge of what we should consider valuable or good. It provides us with knowledge of the good. At this point, we have still not fully explained how we can arrive at specific moral obligations. We have shown how environmental values can be recognized. We can explain that it is good to preserve old-growth forests, to protect the habitat of wild animals and plants, and to prevent the pollution of rivers. What we have not done is explain why we are morally obligated to do certain things and to refrain from others. We must explain how these values we have discovered lead to moral obligation.

Probably, some of you are now impatiently asking what the problem is. If we know what is good, why do we not know what we should do?

Moral philosophers have long been pointing out that we are not obligated to do everything that is good. Some good things may be so difficult or require such extreme self-sacrifice that we hesitate to say that doing them is every person's duty. We might admire people who do such great things, but we consider them saints or moral heroes. In fact we have a term for things that are very good to do but are not any person's duty to do. We call them acts of supererogation, meaning acts beyond what can be asked of a person. If I devoted all my spare time to protecting the Everglades, traveling great distances to talk to people who can help the Everglades and spending every penny I can get in behalf of the Everglades, there would be people who would think me an environmental saint. I might even be compared to Mother Theresa. Would a reasonable person say that such devoted action on behalf of the Everglades is my duty?

The difference between the goodness of doing something and its being a moral duty is not limited to heroic acts of self-sacrifice, however; of all the ordinary, good things a person can do, only some of them are things a person is morally obligated to do. It would be a good thing for me to give money to some of the environmental organizations I do not now support, but it is hard to see this as a moral obligation. How much more am I obligated to give? I will not disagree with the notion that my support of environmental organizations is fulfilling a duty, one I gladly carry out. Will a reasonable person say it is my duty to give money to all organizations, those I supported at one time as well as new ones I have not supported? We can see that a conflict between obligations to my church and some political and social service organizations would soon make it impossible for me to give more to all the good environmental organizations that could use my gifts. One thing we have been pretty consistent about in ethics is that what is impossible is not to be considered a moral duty. There is a point beyond which doing a good thing is not a moral duty. Yes, you should write to your member of Congress, but are you obligated to write every week? Every day? It would be very good of you, but in most cases it would not be a moral duty.

Marking this difference between the good and the right has been a part of the criticism of utilitarianism, which holds that we should do what brings the greatest good to the group of people affected by an action, that we should, in other words, optimize the good. The critics of utilitarianism have pointed out the problems of requiring heroic self-sacrifice and making duties of deeds of supererogation, but they have found another difficulty. If we are obligated to do the most good, we might find ourselves obligated to do unjust deeds that would bring more pleasure than pain to the group affected, in spite of the injustice to a few people.

Discovering what is good does not seem to lead directly to knowing what we are obligated to do. Some moral philosophers treat this move from moral values to moral obligations, from the good to an 'ought', as a very difficult problem.[1]

This supposed logical difficulty has been seen in the difference between ethical descriptivists, who attempt to give a factual basis for value by defining moral value in descriptive terms (such as good being the pleasant or what people desire), and prescriptivists or imperativists, who relate value to obligation by defining moral value prescriptively, by saying that to call something morally good is to say that it should be done.

The descriptivists have taken care of justifying moral value, explaining why the valued thing is indeed good, but they have not shown that realizing the value is a moral obligation. An example of this would be the utilitarian who shows that bow-and-arrow hunting is bad because it causes pain to the hunted animals, but the utilitarian has not given a logical demonstration, it is said, of the obligation not to hunt and cause pain to the animals.

What the descriptivists have done is define value descriptively: the value in question is defined in naturalistic terms, such as the pain of hunted animals or the loss of endangered species in draining a swamp. What the descriptivist fails to do is make the connection between value and obligation.

The prescriptivists[2] approach ethical obligation by defining value prescriptively; to say that an action has value is to say that we are obligated to do it. This relates value to obligation, but the values have been given no factual justification; values and basic moral beliefs about what is good are simply adopted without a factual reason for doing so. It is for this reason that prescriptivists are also called neutralists. Goodness and moral value are neutral, in that they do not require any particular view about the world and human nature.

Now it might seem that we could solve our problem by combining descriptivism and prescriptivism. Logical objection is raised to this, however, because it is seen as an equivocation in the use of the term *value*, which is defined differently in the two ethical approaches. It appears that there is no logical way to relate factual information and obligation. If moral good is understood in factual terms, to claim that something has value is to do no more than to describe it. If moral value is understood prescriptively, to say that something is valuable is to do no more than to urge the hearer to seek it. In the first case, there is no logical connection with reasons for acting; in the second case, there is no factual justification for the value judgment.[3]

Is there any way to avoid this problem and bring together factual knowledge about the world and moral obligation? I believe there is a way. To achieve this, an ethic informed by ecology must treat moral value in a way does not lead to either descriptivism or to prescriptivism (neutralism). Can ecological ethics relate factual information to moral obligation without equivocating in its definition and employment of value? I believe it can do this, but older concepts of the nature of moral reasoning cannot serve our purpose, and the theory of human action implicit in the interpretation of the relationship between *is* and *ought* that we have been talking about puts a logical wedge between approval of a course of action and being obligated to act on what is approved. We can distinguish between approving an action and being obligated to do it. We can distinguish between the good and moral duty. Examples of the difference have been already given. I do not think, however, we should see it as a logical wedge. It would involve a logical wedge only if we thought it was necessary to unite moral value and moral duty by a logical argument. My approach does not seek a logical connection, which seems to be the wrong place to look.

When we examine what actually happens when we reason morally, we find that the connection between the recognition of the good and reasons for acting is immediate. It is not a matter of logical argument. It is not a matter of going step by step from premises to a conclusion entailed in the premises. This has been explained in several ways.[4]

When an environmentalist recognizes that the preservation of a natural area is good, that it accomplishes goals that are morally right, this recognition gives reason enough for the action. The recognition gives a purpose, a cause, and it would strike the environmentalist as absurd to wonder if there was a moral reason to act.

When we examine actual moral thinking, we do not find a logical gap between valuing and reasons for acting. We do not need to justify logically the move from judgment of value to action. The gap some people have seen comes from an effort to connect *is* and *ought* through deductive argument, but such gaps are not there in an examination of actual cases of justified action.

The kind of moral thinking and moral action I am talking about involves neither descriptivism nor neutralism, even though it has some similarity to both and might be mistaken for one or the other or for an attempt to merge them. There is a limited similarity between ecological ethics and descriptivism, because moral value is closely related to factual knowledge in both views, but the nature of the relationship is significantly different in these two approaches. With descriptivism values are thought to be derived from factual knowledge. In my analysis, however, facts are not the logical foundations of values. The basis of value judgments is present from the start in the same lived world that is the basis for factual judgments. There is no derivation of value from facts. There is no need for something like the descriptivist justification of values. Likewise, the similarity to prescriptivism (neutralism) is superficial. Values are seen in my analysis to be reasons for acting, but the reason for this is quite different from the reason in prescriptivism. I do not employ the neutralist tact of treating values and moral principles as action demanding by definition. Value and volition are united in our primitive constitutions of a lived world, just as the materials from which facts and values can be abstracted are fused. The action-guiding aspect of value does not rest on a definition of value in terms of action.

There is no need to combine descriptivism and neutralism in order to frame an argument; not only would this be equivocal, but it would be beside the point, because both views are inadequate and there is no need for a logical argument to bring together factual information, value, and reason for acting. A logical separation of two aspects of value, value as justified by factual considerations and value as action-guiding, is not in accord with actual moral experience. In the making of moral judgments, there is no necessary separation between the descriptive aspect of valuing and value as action guiding. Values are by their nature inducements to action. When values are at stake, there is reason for action. Rather than take the neutralist position, I appeal to the role values and basic principles actually play in decisions about how to act.

The basic moral principles one holds and the moral values one recognizes affect the way a person constitutes the world; an important part of that constitution is what appears desirable and what appears to be called for. Our original experience of our lived world includes a fusion of fact and value and something more, our aversion to some things and our attraction toward other things. The primitive fusion of fact and value provides the substance of our willing, which can be drawn out as conscious desire. There is in this earliest consciousness of the world the material that will become the factual, the evaluative, and the volitional aspects of our world views.[5]

Fittingness and Moral Obligation

Let us examine more thoroughly the way we consciously reflect upon the relationship between awareness of value and obligation to act. When we become consciously aware of our beliefs and think about what we should do, or when we make a judgment about what we have already done, we can see a distinctive relationship between the beliefs and the actions we are obliged to do. The best understanding I have found of that relationship is that it is the relationship of *fittingness*. For a long time, moral philosophers have recognized the role of fittingness; it has often played a role in ethical theory. Understandably, a careful analysis of fittingness has seldom been made; it is such a basic notion that it is easier to rely on the reader's recognition than to explain the concept. Maurice Mandelbaum's description of fittingness, in his *The Phenomenology of Moral Experience*, is the most adequate I know. Mandelbaum does not think that fittingness can be defined, but he holds that it is "a phenomenally objective relational characteristic" that can be indicated ostensively. It can be pointed out in many specific cases and discovered in contexts with which the reader will be familiar. Mandelbaum calls fittingness a relationship between an action and "the initial conditions which call forth an action," conditions that include "past and future events which we recognize as being relevant to the choice we are to make" and a state of affairs we believe can be realized. He cites as an example a person's having made a promise and being able to keep the promise. Mandelbaum says that instances of fittingness can be described as cases of "being in harmony with," "leading to," "completing," "fulfilling," or "answering to." Doing what one has promised can usually be seen as being in harmony with, answering to, and fulfilling the making of the promise.[6]

Abraham Maslow provides further elucidation of the concept of fittingness. He finds explanations in Gestalt psychology of the way a situation demands for its fulfillment certain specific actions. He cites cases he sees as examples of how "poor gestalten make themselves into better gestalten," as when "an incomplete series calls for a good completion." He speaks of how a "musical progression demands the right chord" and how "an unfinished problem points inexorably to its proper solution." Maslow refers to a situation's requiring a conclusion as the "dynamic" or "vectoral" quality of facts. He says this relationship between a situation and the fitting conclusion bridges the fact/value impasse.[7]

When my friend Ka Tresca was studying French, she experienced the way belief can affect fittingness. She was invited to have Thanksgiving dinner with a French Canadian family. Ka's family celebrated the holiday with turkey and cornbread dressing. The light brown dish her new friends served with turkey was pale and bland and soft, really not a pleasure to eat. Later in the meal she heard the dish called mashed potatoes. On again trying the potatoes mashed with the skins on, Ka found it better than the white mashed potatoes she knew at home. The dish she could not eat as turkey dressing was excellent as mashed potatoes. Taste and texture were fitting for mashed potatoes.

It might be argued that moral fittingness would be so different from other examples of fittingness that examples of fittingness in respect to food, the improving of gestalten, different sorts of series, and musical progressions do not show anything important about moral fittingness. Is moral fittingness distinctly different from other types of fittingness? Maurice Mandelbaum holds that there is no distinct moral fittingness.[8] I agree with Mandelbaum that there is not a distinct moral fittingness. I believe moral fittingness is best understood as fittingness between a situation and an appropriate action in which a moral issue is involved. Fittingness in a moral situation might have qualities not present in nonmoral situations, but that is because a matter of moral importance is involved in the situation; it is not an indication of a unique form of fittingness, moral fittingness. Mandelbaum's general descriptions of fittingness and Maslow's examples elucidate the notion of moral fittingness as well as other sorts of fittingness.

It may be questioned whether recognition of moral fittingness is sufficiently transcultural to play the important role in ethics I am giving it. There will be differences in what people find fitting, but the differences may be as strong within a culture as between cultures. I also think there will be more transcultural agreement regarding moral fittingness than with other forms of fittingness. Despite the great differences over fitting table manners, disapproval of murder and theft is almost universal. Can we expect people in most cultures who come to find value in the natural world to see protection of that world as fitting? It would, I think, be most surprising if people could hold the belief and see destruction of the natural environment as fitting. In fact, many beliefs and practices that are favorable to the environment have been present in very different cultures. It has been pointed out that people in cultures with beliefs that would seem to favor the environment engage in practices destructive to it. Need for food, fuel, and water can lead people to do things they do not like to do. What is fitting in one respect, duty to one's children, may not be fitting in another. Should we be surprised when duty to the children wins out over saving the trees?

We do not find a logical gap between what is good and what is the right thing to do. Values and moral principles affect conduct through the agent's constitution of a world; the person sees the world as a place in which some things should be and others should not be, some things should happen and other things not happen. If the values and principles a person professes do not affect the person's behavior, we think something has gone wrong. What a person believes to be morally desirable will ordinarily cause that person to chose actions that accord with the beliefs. A person tends to seek what is valued and to seek the more highly valued in preference to the less valued. If we value something highly, we will seek it unless we have good reason not to do so, such as the item costing too much or causing too much pain to ourselves or others. We have to weigh values against alternatives, and a number of small values might seem a better choice than one valuable thing for which we will pay dearly. I said we tend to seek what we value, because we can avoid seeking something we value. The point is that valuing something is our justification for seeking it unless there is some impediment, such as high cost or too much hurt. We do not need a logical argument to justify seeking what we value.

Various counterforces can cause the behavior to be contrary to the person's professed values or principles. Some people have a low level of moral commitment; other values also held by the agent may override the moral values. Some people have a low level of energy or a tendency to be distracted; they simply are not very good as getting what they want or being what they want to be.

Pressure from other people may keep a person's behavior from being consistent with the person's own values. Some people are easily confused or easily misled, and some people do not have a well-organized and clearly understood system of values. Nonetheless, values and moral principles do tend to guide action, and when they do not the reason can often be discovered.

Of course, some people profess values falsely, and some people can rationalize their behavior to make themselves feel better or to try to make others have a better opinion of them. This seems to have been happening for ages, and there seems to be no way to prevent it, but this does not show that our ethical theories are mistaken. It is a mistake to criticize an ethical approach on the grounds that people might use it dishonestly. The moral authoritarian might claim that it is a weakness of contextual approaches that they allow people to justify to themselves their wrong behavior and act in their own selfish interest. Actually, people have misused authoritarian systems as well as contextual ones; the point I would make, however, is that this attack on nonauthoritarian ethical systems reflects a lack of understanding of the nature of morality and of the purpose of ethical systems. Obedience to authority and morality are different matters, as Kant and many others have made clear; the good will must be autonomous, and the morally right act must be a free act. The purpose of ethics is to guide good people. It cannot make people good nor can it make bad people do the good. We have law and the pressure of public opinion to make people do the right thing, if doing wrong will affect other people significantly. We keep people from murdering, stealing, and various other harmful behaviors; we need protection from these things, but we do not rely on ethical systems alone for this purpose.

The reliance on fittingness to justify moral choices might seem to leave us with an entirely subjectivistic ethical theory. To see why this is not the case, it is necessary to consider the important role of an individual person's world view in establishing what is seen as fitting and to examine various criteria on the basis of which world views can be criticized. The ethical theory I am presenting does not support the notion that ethical positions are immune to criticism. One of the most important kinds of criticism is showing that the world view that underlies a judgment is inadequate. The criticism of world views was treated in some detail in the preceding chapter.

Summary of Theoretical Issues

Because of human weakness and human wickedness, values do not necessarily lead to inescapable moral imperatives, and we are sometimes influenced by conflicting values. I have argued that an ethical system should not be expected to correct the weakness and the wickedness. Later I will argue that our having to confront conflicting values is not damaging to our ethical approach. We have no reason to claim more for our ethics than the approach can deliver or more than it is reasonable to expect of ethics. If we are honest with ourselves and are willing to learn and use our powers of moral judgment, our contextual ethical approach will guide us as well as or better than any alternative we know.

The logical basis I propose for ecological ethics does not try to support a foundation for ethical norms that can be supported by deductive argument, nor does it see the impossibilty of doing so as a barrier to a significant relationship between factual knowledge and moral obligation. The *is/ought* impasse does not prevent such a relationship, unless deduction is the means used to demonstrate the relationship between knowledge and obligation. Turning to insights from theory of perception, the present approach finds in the role of a person's individual world view in the process of constitution of experience the explanation of how knowledge is related to a person's sense of moral obligation. The nature of situations that require us to choose one action over another is given in the act of consciousness in constituting that situation. Knowledge and belief about the world, along with previous judgments of value and other cumulations from previous experience, are important elements in the constitution of the lived world. Holmes Rolston, III uses a striking figure of speech in describing a negative aspect of the connection of belief and moral sensibility. He says that the belief that nature has no normative structures "sinks down into the hinterlands of our minds, an invisible persuader which silently shapes an ethic."[9]

There is a fusion of fact and value in the constitution of a lived world, the only world we can experience. Apart from consciousness, the notions of fact and of value are not clear. When we consider our original experience of the world, we are down to the bedrock of knowledge. It is significant that our primitive lived world is not a world of facts on which we can base value judgments. It is a world in which the matter on which factual judgments can be based and the matter on which value judgments can be based is fused. Facts and values are drawn from this rich ore of original experience. The values may be said to emerge with the facts. There is no need to move from *is* to *ought*. In our constitution of the world, there are no brute facts. There is no pure objectivity or pure subjectivity. When we constitute a state of affairs, the factual aspects of it and the valuational aspects are there together. We see what is, and usually at the same time we have positive or negative attitudes toward it. We place a mode of value on it; it is valuable, indifferent, or something that should not be.

When we have a positive attitude toward something, we have a reason for acting. We have a reason to bring about a state of affairs or to protect what is already achieved. We have a reason to prevent or attempt to change those things we value negatively.

We recognize that some attitudes and some behaviors are called for by the situations we constitute. There is a fittingness some actions have in certain situations. Fittingness gives us more than reasons for acting. It shows us the difference between right and wrong reasons and actions.

We do not deduce the right behavior toward the environment from ecological knowledge, but our understanding of the right behavior toward the environment is influenced by ecological knowledge and by environmental philosophy. This influence, or "feedback," as Holmes Rolston called it, is not a superficial connection between ecological knowledge and behavior. It is not a deduction of duty from factual knowledge; it is an influence of knowledge about the natural environment and of philosophical understanding of the place of humans in nature that happens at the very inception of moral beliefs in the constitution of our lived worlds.

CHAPTER SEVEN

The Status of Values in Nature

Problems in the Language of Value

Discussions of environmental ethics are often couched in terms of value. In Chapter Three we saw that environmental holism can be based on the concept of values in the natural environment. J. Baird Callicott has written on the value of nonhuman species and on the concept of intrinsic value in environmental ethics.[1] Holmes Rolston, III has written extensively on the issue of values in nature and on the several kinds of value.[2] Some of the controversy about the role of value in environmental ethics is indicated in the exchange between Eric Katz and Anthony Weston in *Environmental Ethics* 10.3 (Fall 1988) over whether environmental ethics has been based on the claim that nature has intrinsic value.

One source of controversy and a frequent concern in the literature on values in nature is whether values in nature can be intrinsic or inherent. Sometimes it is asked whether they can be objective. I think the real issue here is whether values that some people see in the ecosphere are somehow really there in some way. Are these values somehow independent of the valuer's mind? Is there something about nature that makes it reasonable to recognize these values? My position is that the values about which environmental philosophers are concerned are indeed real; they are not a result of wishful thinking nor a product of sentimental fancies.

Unfortunately the terms *intrinsic* and *inherent*, which are most often used to talk about this issue, are used in confusing ways. Some writers make a distinction between the two terms, while some use just one term, with one person's *intrinsic* seeming to mean what another means by *inherent*. Some writers try to stipulate a meaning for one of these terms, which is the responsible thing to do, but none of these stipulations become common usage. The dictionaries are not much help. They define both terms in about the same way, often using one to define the other. The only uniform thing to be observed about the two words is that intrinsic and inherent value are contrasted with extrinsic value and instrumental value.

It has seemed to many philosophers to be clear and unproblematical to say that natural values are objective or to deny that they are objective. The concept of objectivity has philosophical problems, however, and the term *objective* is too broad to convey the precise basis for holding that a value is solidly grounded and worthy of being held. For these reasons, I believe we can be much clearer and avoid problems with the way we talk about value.

Some developments in modern thought, especially in subatomic physics and in our growing understanding of perception, indicate that we were simplistic in referring to values as either objective or subjective. Recognition of value, as well as all our perceptions of objects in the world, has an aspect of subjectivity, in that the conscious human subject contributes to the perception, and an aspect objectivity, in that an extra-mental world contributes to the perception. We can now see that the division of ideas, values, facts, etc. into those that are subjective and those that are objective is an inadequate way to understand them.

Even though it has generally been taken for granted in Western thought, especially since Descartes, it is now time to abandon the comfortable old dichotomy of knowing subjects and objective knowledge. Developments in contemporary science and philosophy show that there is no factual knowledge that is a product of the world alone, with no influence from the knowing subject; also, there are no mental things that owe no debt to the extramental world.

One development in modern thought is contemporary physics, with the relativity principle, the principle of complementarity, and the indeterminacy (or uncertainty) principle. The old way of thinking about the knower and the known simply does not fit what is happening in quantum mechanics. Several philosophers have pointed out implications of the breakdown of the subject/object dichotomy in physics. J. Baird Callicott has written of the significance of this for ethics and value theory. He argues that the end of the subject/object dichotomy supports the ascription of value to the ecosystem.[3]

An objection might be made to this, holding that what applies to the microlevel of quantum mechanics does not clearly apply to the level of ordinary experience. This objection does not follow, however, because contemporary philosophy of perception, dealing primarily with the macro level of ordinary experience, requires the rejection of any subject/object dichotomy. As we saw in Chapter Five, all awareness of objects is an activity of consciousness, known technically as constitution. Consciousness is always active in perception, attending to some object, and it places an identity or thesis on the object to which consciousness is attending. This is not to say, of course, that the species of northern orioles underwent a change in itself when the authorities on the subject decided that the spotbreasted orioles are not a separate species, but are northern orioles. The denizens of the abysmal depths did not change in themselves when we found out about them. The point is that things become things of specific sorts within our lived worlds when consciousness makes them things for us within our lived worlds, and that is important because our lived worlds are the only worlds to which we have access. Without the constituting activity of consciousness, we would not have this lived world, the world we are aware of and in which we carry out the various activities of life. This lived world is the world which we are talking about when we speak of the real world. What this entails is that nothing is objective in the sense of being independent of the consciousness of the subject who is aware of it and who had a role in its being known with those qualities we perceive in it. Everything comes to us through our subjectivity, our active and constituting consciousness, from the microlevel activities of electrons to the affairs of ordinary experience. This is not a simple coming through, as the scene outdoors comes to us through the clear glass of a window. This activity of consciousness involves culture, often including the cultural achievement of scientific theories, as well as the physiology of our nervous systems.

As we saw in Chapter Five, we would not be warranted in jumping to the conclusion that all the objects of which we are aware are merely subjective and unrelated to any extramental world. It is a mistake to consider objects in the world simply objective, but it is also a mistake to think of them as merely subjective. The terms *objective* and *subjective* need to be replaced with more precise and informative language. We really do not need those terms and can replace them with descriptive terms, such as *well founded*, *based on strong evidence*, *widely experienced*, *generally agreed upon*, and *intersubjectively verified*, along with other expressions that indicate the precise grounds on which a statement is supported. At times we can say who did the experiment, when and where, what happened, and what it tells us. In contrast we can use such terms as *imaginary*, *not supported by evidence*, *biased*, *wishful thinking*, *mere conjecture*, and other expressions that not only announce that we should not accept some claim to know something but also explain why a claim is not worthy of acceptance. Saying that a statement is subjective does not tell us as much as saying that the statement lacks the agreement of most careful observers, that the person making the claim lacks competence in the field, or some other really informative observation about the claim in question.

We realize that some things are fanciful and that other things are substantial and dependable parts of our daily lives. We realize that some beliefs are based on unconfirmed reports and others are grounded in the experience of diligent observers. We know that some perceptions come upon careful observation and others rest on halfhearted attention. These distinctions are real and important. We do not think adequately or live well without making them. We do not explain our critical judgments about perceptions and beliefs very well, however, when we talk about objective beliefs and subjective beliefs. I have indicated two reasons why this is so. First, the concept of perception suggested by talk of objectivity and subjectivity is no longer supportable. We do well to abandon notions of perceiving and thinking that have been abandoned for very good reasons. The second reason to stop talking about perceptions or ideas as subjective and objective is the value of using expressions that are clearer and more precise.

Our thinking that some things are more objective than others can be better understood as the realization that some things have more intersubjective support than others. There are some things virtually everyone, and especially everyone in our culture, constitutes the same way. This does not make our knowledge of tables and chairs "objective." We all recognize the common sorts of furniture and tools and food items used in our common culture; this is the result of intersubjectivity, which is the closest thing we have to objectivity.

I suggest that we would do well to stop talking about ideas or beliefs being subjective and objective. Philosophical use of these terms must be limited and more technical than the ways we have most frequently used them. The important thing is that we recognize that we must take into account two factors: (1) the working of consciousness (which is the subjective aspect of our perceiving a world and thinking about it) and (2) the extramental world itself (which is the nonsubjective aspect). The extramental world comes to our awareness only through consciousness, and this occurs in many different ways, some of which give us good reason to accept the reports of experience and the beliefs which are compatible with these reports. All things in our lived worlds are constitutions of our consciousness, but they are not all merely whimsies and fancies of our minds. We observe most of what we see in the world because there is a world to be seen, and we find that our fellow humans are seeing the same things we see. We have a number of ways of judging the claims people make on the basis of experience. Part of what I am urging is that the way we talk about our reasons for rejecting one belief, treating another with great caution, and feeling confident about another not be reduced to the vague and confused notions of subjectivity and objectivity. We can say precisely why we reject astrology, have doubts about mental telepathy, and accept with confidence studies of the effect of mercury on the streams and lakes of south Florida and on the people and panthers who eat the large mouth bass.

A Taxonomy of Values

In order to make a taxonomy of value words I think could facilitate discussion of the basic issues in value theory, I proposed the use of some new terms (free of connotations of older terms) along with the use of older terms that are not problematic.[4] Even though there does not seem to be a mighty change in usage following my article, I will follow my own taxonomy. I regret the awkward sound of some of my terms, but I hope the awkwardness itself will have the advantage of not leading to confusion with more familiar terms and the notions associated with them.

One type of value is that of a thing valued for itself; it is referred to as an end in itself. The qualities or entities that have value in this way are not valued primarily for what they can achieve as means or as instruments for securing something that people want. This value as an end in itself I call 'end-value.' What is meant by "*x* has end-value" is "some *s* values *x* as an end in itself."

There seems to be no limit on the kinds of things that can have end-value. All sorts of objects and qualities can have end-value if someone values one of them for itself. Some people value trinkets and found objects that other people consider worthless. Some people value personal qualities such as health or fame or virtue. Some people value things that loved ones made or once possessed. All it takes to ascribe end-value is valuing the thing for itself. This means that end-values do not need to be justified. Some end-values can be very important things, and some of them can be justified, as when someone values the alligators in the Everglades just because that person likes them. There are better reasons to value the alligators, such as the role they play in the ecosystem, and the person who just likes them may one day learn just how important they are, but insofar as end-value is concerned, no reasons need be given. That is why I feel justified in keeping some small rocks I picked up years ago.

With end-value, having value is the same as being valued. This may be the case with all types of value, as many people think. If having value is the same as being valued, talk of a thing having value is a way to talk about that thing being valued by someone. Why do I talk about value, using the noun, as though it referred to some quality in an object? Convenience is one reason for using the noun *value* as a sort of shorthand for speaking of a person valuing something. Another reason to continue using the word is that the question of whether all value exists only in the valuing is yet to be discussed. I do not want to accept a way of speaking that will prejudice further inquiry. I am not convinced that all value exists just by being valued, at least not in the direct and simple way some of my colleagues connect value with being valued.

If there are any values that exist independently of being valued by someone, we need a term for them. I will use 'independent value' to designate values that are independent of human valuing. What is meant by "*x* has independent value" is "*x* has value independently of being valued by some *s*." If *x* has independent value, speaking of the value of *x* is not a shorthand way to talk about someone valuing *x*, and the value of *x* is not derived from the fact that some person values *x* or may come to value *x*. This would mean that forests and rivers, a diversity of species, and some of the other things we value in nature have their value apart from our thinking them valuable.

Independent value is the kind most likely to be questioned or immediately denied. Some writers say that the notion of value apart from a valuer has no meaning for them. The very notion of independent value confronts a well-entrenched dogma, which may or may not be justified.

If the concept of value that does not depend on human valuation is controversial and rejected by many, possibly most, philosophers, what reason have we for examining the concept of independent value? Have we any evidence to suggest that such value exists? One bit of evidence is that we now value some things that were once not valued or were even unknown. Speaking of "presubjective value," over a decade ago Holmes Rolston, III called attention to such valued things in nature as myoglobin, from which hemoglobin evolved; structures in the ecosphere such as nutrients in food chains, and processes such as glycolysis and photosynthesis.[5] In his more recent work he has developed more fully his defense of the independent status of values in nature.[6] I know of no better examples of what I call independent values in nature than those given by Rolston.

It seems a bit odd to think that the marvelous parts of our world, including the parts of our bodies, suddenly became valuable when people learned about them. Did my islets of Langerhans become valuable that day in biology class? Did the plants in the oceans that make most of the oxygen in our atmostphere only become valuable when humans finally discovered why we have breathable air? Do we not sometimes discover that something has value and had it even before we knew it? Rolston thinks it is arbitrary to deny that there can be any value that is independent of human valuers, and it seems to me he is right. The important matter is to examine values that do not seem to exist through our valuing and to determine what sorts of things they are.

The concept of virtual value, which is a capacity some things have to be valued when someone discovers them, does not seem to explain very much. It leaves the difficult questions unanswered. It strikes me as a way to preserve the dogma that all value consists in being valued, and this dogma needs to be examined, not preserved through the use of notions that might be question begging. It might turn out that there is no value apart from a valuer, or that the values that can be explained as independent values are not very significant. On the other hand, the concept of independent value might be quite significant. It remains to be seen.

In addition to exploring the notions of nonextrinsic value, end-value, and independent value, we need to examine some of the ways a thing can be valued as a means. The notion of instrumental value includes too many ways of valuing. It needs to be divided in order to facilitate examination and allow for the making of important distinctions.

There is an important difference between the value of a tool and the value that something can have as part of a system. One way of seeing this is to note the ease with which one tool can often be replaced with another. Changing parts, however, can change the nature of a machine, an organism, or a system quite significantly. The very identity of a whole might be called into question if a certain part is replaced. There may be cases in which certain tools are necessary for the functioning of a system, so that their being just tools or part of the system might be debated, but usually the distinction between the instrumental value of a tool and the value of a part can be maintained. There are a number of ways a part can be valuable to a whole. Any part of an artwork may be essential to the work, a character in a play may make the play, and even an oboe may be necessary to give a musical composition the character for which it is known. The value of being a functional part of a whole natural system is the sort of value on which I wish to focus. This is a kind of value that has, I believe, special significance for environmental philosophy.

In order to talk about the value something has as part of a system, I will use the term *functional value*. "X has functional value" means "*x* has value as a significant part of a system (which is of the sort valued by virtually everyone, or would be valued if known)." Holmes Rolston's examples are of things that have functional value. There is a serious question regarding the definition of functional value: should the part in parentheses about the system's being valued be included? It may be unlikely that we would ascribe value to the parts of a system we do not value, but we might need to acknowledge that these things have value for the system, whether we place any value on the system or not. We will need to examine this issue later.

Combinations of Value

It is possible for one entity to be valued in more than one way. A person may value something as a tool while also valuing it as an end in itself. What is the significance of this? Is the possession of more than one kind of value simply fortuitous, or are there some logical connections between the several kinds of value? Is the possession of certain pairs of values logically or factually impossible?

The combination of values I believe to be of greatest potential interest is one entity having at the same time both end-value and independent value. If one thing had both these types of value, that thing would be valued for itself alone, not as a means to something else. The value that thing had would not depend on the fact that it was valued; it would have the value even if its value were not recognized. This combination would be, I believe, what most people are asking about when they ask whether values can be objective. What is the connection between the combination of end-value with independent value and the notion of objective value? What are people looking for when they seek objective value? It seems they are seeking something that is not instrumentally valuable. This seems clear from the repeated contrasting of intrumental value and 'intrinsic' and 'inherent' value. The sort of values that are considered objective would have to be end-values. Often it appears that being an end-value is all that objectivity, found in 'intrinsic' value, would require. My examination of end-value, a thing being valued as an end in itself, indicates that this kind of value is not enough because it can be quite arbitrary. It seems very important that objective value not be arbitrary. What could keep a valuation from being arbitrary would be its independence from the fact that some person values the object.

All this talk of end-value and independent value might seem complicated, but it would be a clearer way of putting the concern over objective value because it avoids the conceptual confusion of the term *objective.* Talk of objective value often elicits immediate denial or fervent avowal, both of which interfere with good discussion of a difficult issue. Perhaps my awkward terminology will make discussion more profitable, even if cumbersome.

Is it possible for one entity to have these two sorts of value, independent and end-value? I see no reason why something could not be a bearer of independent value and be valued as an end in itself. The more important question is what significance this would have. Would this be of any importance, or would it be a fortuitous conjunction of no real significance? Before we look at examples of things with both types of value and attempt to assess their importance, let us be sure that we are clear about the nature of the types of value involved.

What makes end-values valuable? It seems to be no more than the fact that they are valued. In spite of the significance sometimes seen in something being valued as an end in itself, to talk about something having end-valued can be misleading. We may be saying nothing about the thing itself, only something about the person doing the valuing. This does not, however, seem to be what people have in mind when they say that something is valuable. To say that an object has value seems to be saying that it possesses some quality of its own on the basis of which some person values it. People who ascribe end-value to an object seem to intend to say something about the object, not only about themselves. When asked to explain what they are saying about an object when they call it valuable in itself, the valuers frequently refer to qualities that are not reasons why other people would value the object. There is usually something about the object on the basis of which it is valued, but it might be something of no significance to anyone except the valuer. When we clear up the semantics of end-value, we see that an object is valued because of a quality of the object, but the quality may be valuable only to the valuer and only because the valuer values it. It seems that ascription of end-value is talking about an activity of the valuer, not significantly of the object valued.

Perhaps I am wrong in seeing end-value as an activity of the valuer. Maybe there is more to it than that, but when we try to investigate this matter, we run into obvious difficulty. What are we looking for, and how will we recognize it? Even if we found something that almost everyone values for itself, what would we have found? We would not have demonstrated any more than the popularity of the thing. The popularity gives us some reason to think there might be something about the thing that makes it something to be valued, but it does not identify that reason for ascribing value to the thing. In their work in aesthetics on the role of critical reasons in the criticism of artworks, Monroe C. Beardsley, Arnold Isenberg, and Paul Ziff make distinctions between types of critical reasons. They explain this in slightly different ways, but the basic distinction is between reasons for supposing that an artwork is good and reasons that explain why the artwork is good. Knowing that the work was done by an acclaimed artist during the artist's most productive period, or knowing that the work is displayed in a major museum, gives us reasons to expect the work to be good. This is what the popularity of a value does for us. Knowing what there is about an artwork that makes it good provides us with the most important type of critical reason, such as the unity of the work or its power of expression. The popularity of a value does not give us a comparable kind of reason for valuing an object. What could show us what there is about a thing that provides us a reason for valuing it? What could show us wherein lies the value of an object?

There have been efforts to show that this or that end-value has an important ground. Most often this has involved a metaphysical or theological answer, but, because the metaphysics or the theology are not as widely accepted as the value, no great agreement has been generated. The values about which a broad consensus is discovered seem actually to be valued as means. A few people may value life because they believe God has given it to us, but I think most people value life because it is the only means we have to experience anything at all. Most people who value beauty seem to do so because it fulfills a function, that of giving an experience most people enjoy.

Claims that life and beauty are simply valuable in themselves are not very convincing. If being dead did not keep us from having the experiences we can have while alive, being dead might be a good idea. It would be cheaper, and we could smoke and drink all we want without any health worries. Being more serious about this, consider the advantages of being a ghost. When I was a child I used to think it would be nice being a ghost and being able to know what was going on without being threatened by what is happening. Now I realize that the advantage would be purchased at a dear price. The ghost does not have some experiences we value. What can the ghost realize in a martini or in a turkey dinner? How can a ghost make love, or even share a joke with a child? I had good reason to give up my envy of ghosts as I gave up my belief in them.

But what does this say about the value of life? I think I value life because life enables me to be in the world and enjoy the pleasures, adventures, and satisfactions life can bring. On the other hand, what can we think about being alive without experiencing anything? Would that be valued? Many of us have signed living wills to make sure that such a life does not happen to us.

How do we feel about beauty when it does not give us any aesthetic satisfaction? When we feel quite ill, in pain or nauseated, most of us do not care a bit for beauty. But what about virtue; do we not value virtue for itself? If we try to ignore the cant and hypocrisy that often comes with talk of virtue, can we find at least one virtue we value for itself? When I go through this seach I find virtues I value in certain contexts, but I think of other contexts in which they are destructive or in which they seem to be tainted by too much self-concern. There is no reason why a person cannot value a virtue for itself alone. I think some people I know do that, such as the person who values courage with a passion or the person who makes a fetish of modesty. The placing of high value on some one virtue is usually taken by other people as a peculiarity of the person in question. Even though a person can value a virtue for itself, there does not seem to be much significance to doing this. It does not show that even the virtues ought to be valued as end-values.

When we turn our attention from single virtues to the virtuous life, we find some moral philosophers claim that we should prize the life of virtue for itself alone, that it is morally wrong to look for some point to morality. This claim is far from being accepted universally by moral philosophers, however, and I agree with those who see the moral life as having instrumental value in helping us find deep and abiding satisfaction in life.

What sense do we now make of the claim that we ought to value certain things merely as ends in themselves? I am not sure we can find any such things, and another problem with the attempt to find such things nags me.

Can there be reasons for valuing something as an end in itself? It strikes me as odd to even think about there being a basis for end-values. How can there be reasons for valuing things as ends in themselves? Are the qualities that justify our valuing also valuable? If we could have the qualities without the thing we value, would we be satisfied? I am not sure we can have a reason for ascribing end-value to a thing x without making that thing a means to something else, a separate and different value y, which is the reason for valuing x. In this case x is not merely an end-value. It would be an end-value only if the first value x would be valued even if it did not produce the additional value *y*. Am I valuing the fertility of the soil in my garden merely as an end-value if I have a reason for valuing the fertility, namely the tomatoes and turnip greens that are my reason for working the garden? Of course, we can value for itself alone something that has value as a means, treating its means value as coincidental. I might value a sugar bowl for itself alone, not because it is handy at tea time. I cannot, however, point to its usefulness as the reason why I value it for itself.

To give the demand for objectivity its due, let us explore it further. What the demand for objective values is asking, it seems, is that some things that are valued for themselves, things with end-value, also have that value independently of human valuing. They would not be valued simply as means to something else, and they would be worth valuing. There seems to be no reason to doubt that there are some things that have independent value and are valued by some people as ends in themselves. My questioning is directed to the significance of this recognition of independent value joined with ascription of end-value. If someone learns about the role of alligators in the Everglades ecosystem, the person recognizes the independent value of these animals. Specifically, that person realizes they have what I designate as functional value. If the person later comes to attribute end-value to the alligators, does that person add anything to the value of the alligators? It seems to me that the alligators have no more and no less value than they have had for centuries. Yet this valuing as ends in themselves, along with the recognition of independent value, seems to be what people have been trying to talk about when they spoke of objective value.

In the alligators, in many other animals and plants, and, I would say, a fortiori in whole well-functioning ecosystems, we have what people have considered objective value. We have things that are valued for themselves and not for any use they have for someone, and they are things that have value whether anyone values them or not.

I have already explained why I think it is a mistake to use the term *objective* to designate values, but I want to probe more deeply. My concern is more than a verbal matter. I am concerned about the significance of something having independent value and also being valued for itself (end-value).

Whether the combination of end-value and independent value is as important as it might seem at first sight, it seems to be operative in the thinking of many people. It seems to provide the kind of values that seem to give a sound basis for environmental ethics. It takes a high level of sophistication to base one's morality on values and not be influenced by the concept of something that is not a mere means to something else and also ought to be valued. A champion of civil rights might speak of the intrinsic value of an individual human life. Is this a turning to end-value, to independent value, or to both? Because end-value can be quite whimsical, and independent value might be instrumental, it seems that both end-value and independent value are implied. The environmentalist who appeals to the value of a stable ecosystem is not talking about a whim but about something that should be valued by everyone and that would have independent value. This sort of appeal is made to support moral claims that are believed to be independent of human interests, human tastes, or social norms.

Can there be any objectivity in this sense? There do not seem to be any logical reasons why a thing cannot have end-value and independent value. We have seen examples of things that might have independent value and be valued for themselves alone. My question is whether this would be a significant combination of value and valuation. Would it increase the value if a person ascribed end-value to one of the things Holmes Rolston, III described as having "pre-subjective value"? There is nothing to keep a person from valuing glycolysis as an end in itself, not because of an interest in metabolism. This might be a bit unusual, but it is possible, so long as the person did not try to justify the ascription of end-value by reference to the independent value. It is important to see that even though recognition that a thing has independent value might be the psychological cause of a person ascribing end-value to that thing, it cannot be a logical justification. The reason for this is that all the things in natural systems that we have seen as examples of independent value are things that have functional value. They are not valued for their own sake but for the contribution they make to the natural system.

The two concepts of value, independent value and end-value, seem to be separate and logically distinct notions as long as we are talking about values in nature, which is what we are interested in. The combination of these two kinds of value may not have the significance that the notion of objective value seems to require of them. There seems to be no logical connection between end-value and independent value; the ascription of end-value to something with independent value can be utterly arbitrary. End-value and independent value are different concepts, and joining them does not add anything to either of them. The combination does not seem to achieve anything. Someone's ascribing end-value to an object gives no reason to expect it to have independent value, and the presence of independent value does not lead logically to ascription of end value.

So far our argument has rested on the arbitrariness of ascribing end-value. Let us see what we find when we examine independent value more carefully.

How Significant Is Independent Value?

What sorts of things can have independent value? The best candidates I have seen are things that were not valued at some time in the past because people did not know about them. They could be things that were inaccessible, such as "many a gem of purest ray serene" in dark ocean caves or the flower "born to blush unseen, and waste its sweetness on the desert air." Thomas Gray[7] talked of unsighted gems and flowers not smelled in order to praise "the rude forefathers of the hamlet," who were not exactly unknown, if not widely known, and most of them were probably valued by someone, if only their mothers, so the gems and flowers are better examples for our purpose. We could include in our examples a lovely poem lost when a haughty lady would not open her suiter's letter or a sonata lost because the composer feared rejection by the public. Why do we speak of gems and flowers, poems and sonatas as things of value? These are the sort of things we value because of the pleasant experiences they give us when we experience them. The pity of the undiscovered poem and the unperformed sonata is that they never do what they could do. These are things we can recognize as having value because they are good for something, for seeing, hearing, smelling, or selling. More important for environmental ethics are things such as those named by Holmes Rolston, III. What are they like? They are all things that have functional value as parts of a system. They are valuable independently of human ascription of value, even in the absence of any human knowledge of them, because they are important parts of life, environmental, or evolutionary processes. We value these things, once we know of them, because we have discovered their value to systems we recognize as good ones. It can now be seen that these gems, flowers, works of art, and the parts of the natural world have value independent of human evaluation in one sense, but they are very closely connected to human valuation in other senses. Gems, flowers, and artworks are valued for the desirable experiences they give to people. The parts of the natural world have long had functional value as parts of a system we now recognize as valuable.

I do not know how to free gems and poems from their connection to human valuing, but there is another direction we could take with the parts of a natural system. We could drop the requirement that functional value be related to a good system, but there are problems with this approach. It is hard to see how we are justified in including in the concept of independent value things that contribute to systems we do not value. If we agree to let the concept of independent value include things that contribute to any system whatever, even those we do not value, the concept becomes a strange notion of value, because value is a normative and not merely a descriptive notion. If the designation of something as valuable appeals to nothing more than usefulness to a system of dubious worth, do we add anything by calling it a value? We should not be misled here by the way we use the word 'good' to designate a weapon for assassination or something else put to a bad use. We would do well to call such things efficient instead of using a normative term to designate something we do not really value. At least I hope you do not value the use of wire to strangle a person, even though we are told that it is a "very good way" to carry out the dreadful deed.

We seem to be limited to independent values that have some type of instrumental value. The most interesting of these are things that have functional value as parts of a valued system. This usefulness of our independent values presents us with a problem in claiming that it is independent of human valuation. Claiming that something has independent value requires that we identify it as valuable by showing how it is valuable independently of human valuing. All we have been able to show is that at some time the valuable object was not known to an evaluator. We have not been able to show that the basis of its value is especially different from that of the many things we value because they are good for something. There does not seem to be any way to show that a thing is valuable without showing that it contributes to something that is good. It must be good for smelling or looking at or some other kind of enjoyment, or it is valuable because it has a function in a good system, whether humans know about it or not. Even when humans did not know about the system, the system was a good one. Why was it a good system? It was a good system because it had qualities humans value, such as life, dynamic stability, interdependency of many species, and natural recyling of materials in trophic webs. The things and the system that fosters them are good in terms of human evaluation. Unless we are willing to call anything that happens in a natural system good, we cannot escape the element of human evaluation. I think it is a mistake to think of nature as a moral system or as any kind of normative system. I believe we can avoid confusion if we recognize that natural systems have qualities we value and qualities we do not value.[8] Some of the qualities we do not value in themselves serve an important role in systems we consider good, such as pain, which is information in the system, warning of harm to be avoided. Pain, death, and other qualities we do not value have value to the system that has developed and sustained life. We can recognize that the system has produced life, without adopting a teleological interpretation of nature that sees nature as consciously goal-directed. The important thing is that natural systems have produced some good things, such as life and beauty, and various things were instrumentally important in reaching these results. We value these good things, but we should remember that making evaluations is a human activity, however, not an activity of nature.

When we limit functional value to having a good function, that is, a function in a good system, it is not absolutely independent of human valuing. The notion of a good system implies human evaluation, unless we are going to call any natural system whatever good and all qualities in natural systems good. I am not willing to do that; the natural events that are now destroying reefs off the coasts of Southeast Florida are not good; they are natural systems gone awry, whether human waste materials are causing the damage or events that cannot be attributed to humans are causing the destruction of reef life. What can it mean to call a system good except that an evaluator has recognized that it is good, in many cases that it has been good for centuries or eons. This is not as simple, of course, as calling a smell good because an evaluator found some satisfaction in smelling it or calling a thing good to see because an evaluator found satisfaction in seeing it. We are talking about complex systems that have been sustaining life and beauty for a long time. We want to say, we should say, that these systems were good long before we knew about them and evaluated them. None the less, the independence of these independent values in nature is not entirely independent of human standards of evaluation.

If that is not enough to discourage us, it has been suggested to me that we have another difficulty. I do not think this is a problem, so let us put it away with dispatch. The supposed problem is that if we say that *x* has independent value because some *s* values something other than *x*, namely some *y* within which *x* has a significant function, we run into a logical problem. If the only *x*s that have independent value are serving a function in *y*s to which we ascribe value, and are valuable because they share in the value of their *y*s, are we not committing a fallacy of division? The fallacy of division consists of ascribing to a part the qualities of the whole, as we would be doing if we argued that a large forest must consist of large trees. In the present case we are doing something significantly different. We are not ascribing value to a part of the valued system simply because it is a part. Let us be clear about this; the functions to which we attribute independent value are valuable for their contribution to the valued system, not merely their membership in the system. It is not a case of the part sharing the value of the whole but of the part enabling the whole to exist as a valuable system or, in some cases, to exist at all.

We still have to deal with the matter of the limited independence of independent values. To call a system good involves evaluation, which limits the notion of independent value. Can we get around this by arguing that the system in which our things of putative independent value function is a system we ought to value? If we ought to value it, whether we do or not, it would have a value that is not dependent on our valuation.

Perhaps we can make a meaningful and supportable claim that certain things should be valued. It is surely significant that the systems we have in mind when we talk of the functional value of *x* are rich, vital, vast systems basic to human life and all other life. We would not claim that every system with functioning parts should be preserved just because it is a system. There seems to be no wrongdoing involved in ceasing to value some machines. The vastness and inclusiveness of the systems whose parts we value may be qualitatively significant. We are not talking about parts of an automobile or a house. The things identified by Holmes Rolston, III were things on which all life depends. A person who does not value a system on which all life depends is either ignorant or wicked. There are people who lack even an elementary knowledge of ecology, so that they think of human life only in economic or political terms. Such a person may be accused of having self-defeating, irrational values. If a person knows that human beings are a part of nature, dependent on the natural environment, and still does not value those functions that make life possible, we consider this person insane or wicked (guilty of the violation of a sort of categorical imperative). If this is too strong a claim, we can at least point out that we have very good reasons for valuing the system on which life depends. We might be able to go further in claiming that a certain *x* should be valued by noting that, in Rolston's examples, the system of which *x* is a vital part is one that is necessary to the continued successful existence of every valuing subject; value itself depends on the existence of *x*.

Even though we have found such good reasons for valuing an *x* as to make failure to value *x* almost incomprehensible, and certainly reprehensible, we have not found values that are completely free of human valuation. Even though we have seen that values can be identified that are not directly dependent on human evaluation, because they can be valuable even without humans having knowledge of them, these values are dependent on human standards of value. When we discover the value of these things, we have a sense of discovery and recognize they have been valuable before we discovered them, but their value is not independent of the value of the system of which they are important parts. Also, these values are not end-values; we do not value them for themselves but for the function they have in a valued system.

Of course, we could value them for themselves if we were so inclined, but this would not be a significant thing to do. I think most people who realize the value of the sorts of things that Holmes Rolston, III talked about would consider it a strange foible to value them as end-values.

Some important points seem to emerge. First, there is a sense to be given to the notion of independent value that makes it quite different from end-value, but there are some problems with the concept. Second, establishing independent value has not demonstrated that this sort of value is all that was sought in the effort to make values objective. The main purpose many thinkers have had for seeking independent value, which they probably referred to as *intrinsic* or *inherent* value, has not been realized. We have not seen clear, important, and unproblematic examples of independent value that were at the same time end-values, and if we did find them, the significance of this would be questionable. The desire for completely objective values does not seem to have been met. Some writers seem to have thought that valuing something as an end in itself was all that was required for *intrinsic* value, which was taken to have a sort of objective basis. I think my analysis of end-value has demonstrated that valuing something as an end in itself is not really what was sought in objective value. I have done what I can with an analysis of these concepts of value. Now I must leave it to those who might claim that there are objective values to present their case.

I see no way to avoid the conclusion that the question about values being objective must be answered negatively. There seems to be no way to justify the claim that there are objective values. The pursuit of objective values is, I believe, the wrong way to support moral claims.

Beyond Subjective/Objective Values

Is the whole thrust of our investigation negative? Do we work at this issue only to wind up agreeing with the moral skeptic? No, this is not the outcome. There seems to be nothing that we should call objective value, but the lack of objective value is of small consequence for our moral thinking. The lack of objective values does not make value merely subjective.

In Chapter Five we saw that perception of objects in the world involves elements of subjectivity, along with the existence of an extramental world that we can see as the objective aspect. You might have no trouble with this as far as the constitution of objects is concerned, but what about value judgments? We have been taught to think there is a profound difference between knowing objects and holding values. There has been some apparent justification for thinking that knowledge of objects in the world rested on a firmer foundation than judgments of value. What has happened is that we have had more intersubjective verification of our perception of objects than we have had of our impressions of value. It has been easier to get intersubjective verification of the natural or physical significations of things in the world than of the cultural significance of things. Perhaps we have worked harder to get intersubjective verification regarding the things that must have been crucial to us when our lives were constantly threatened. Of course, the natural significations mattered most to us then, and we have been slow to put matters in another perspective.

This more easy access to intersubjective verification of the natural properties of objects has led many people to think the physical aspect of things is more "real" than the cultural aspects, including value. A clearer understanding of perception can show us that this is mistaken.

Our values are constituted the way facts are constituted. In our original exposure to things, in our turning toward things in the world before we have constituted these things as this sort of thing or that sort of thing, facts and values are not distinguished, but together and undifferentiated. The Gestalt psychologist Abraham H. Maslow speaks of them as being fused. They may remain fused in much of our experience. Maslow has given some examples of fusions of fact and value that he had experienced. He tells of his experience with new studies of the use of hypnosis in medicine. As his familiarity with hypnosis grew, he also came to value hypnosis, not because of logical arguments based on the facts he was learning but because of a valuing that was fused with the learning.[9]

In reflection we separate out cognitions, evaluations, and volitions, but we found these originally together in our lived world. Each type of attention to the world is as original as another.

Evaluation does not need to ride piggyback upon cognitions, but the two are related, not because one is derived from the other, but because they are both derived from the same original contact with the world. Volition is not derived from evaluation. The relation between the two is both more intimate and more equal. The particular arrangement of cognition, evaluation, and volition varies, but it is present in all constitutions.[10]

We can abstract or extract facts or values, giving a name to this, a name to that. This requires us to pull away from experience a part of our lived world and to constitute it as something in itself.

We can constitute facts, and we can constitute values. We do not usually constitute facts and then derive values from our facts. Neither fact nor value seems to have any necessary priority. Consider your constitution of a bed of flowers when you came around the corner and there they were. Did you pick out the shapes and colors first, then realize that the flowers were beautiful? Perhaps you came to realize they were tulips while you stood speechless before their loveliness. What do you remember about seeing the tulips? Have you forgotten the exact colors but retained a clear memory of the beauty, of how it made you feel? It is not unusual for the evaluation to come first, and this realization of beauty stands on its own. It does not need to be supported by a description of naturalistic significations. When our experience follows a pattern of acquiring facts, then making an evaluation, and finally deciding what to do, we have engaged in a special case of reflection. This has also been a rather difficult process, involving thinking and deliberation. It is fortunate we do not always have to work things out that way. Sometimes we perceive the value and beauty of an undeveloped piece of land. Immediately we want to protect it. We might engage in a laborious process of gathering scientific data about the piece of land only because we need to convince others that the land should be preserved.

Our constitution of factual data and our constitution of values are not greatly different processes. Coming to know facts does not rest on sturdier legs than evaluation can have. One type of constitution is neither stronger nor weaker than the other. Sometimes we do one well and the other poorly, but neither can claim to be inherently more important or logically prior to the other.

This analysis, which reveals the role of consciousness in factual knowledge and in evaluation, may seem to knock the props out from under all our knowledge and all our values, but that does not happen. The range of intersubjectivity, incorporating facts and values, is quite extensive. We do have something against which to check our perceptions, feelings, and desires.

Another important factor is that our values do not come as isolated individual pieces, any more than our facts come to us in fragmented fashion. How can I value the birds in the Everglades without valuing their source of food and other aspects of the habitat that enables them to live? Values fall into clusters; one value leads to another, calls for another, takes strength from another. Much of our skepticism about values can be attributed to our penchant for examining values individually and subjecting them to a fierce scrutiny. It is no wonder that the value we are examining, like a friendless and helpless suspect tied naked to the rack, makes a rather poor showing. We know not to treat our facts this way, and we would do well not to treat values in such an unfair way. Values are no more independent of other values than facts are of other facts.

The outcome of this analysis is that we have made a serious mistake in thinking of knowledge as objective and value as subjective. Value and knowledge are on all fours together. This need not make us think less of knowledge. What it does is indicate that we can treat values in much the same way we have treated knowledge. We can make judgments, based on consistency and intersubjectivity, between knowledge claims and value claims. We can, in some significant ways, reason about values.

Can values be objective? No, it appears they cannot. Does this matter? Does it undercut our efforts to make responsible ethical judgments? No, it does not have that effect. In fact, it may turn out to be a real advantage. Leaving behind the subject/object dichotomy, we may be able to understand ourselves and our world as we were never able to do before.

CHAPTER EIGHT

Contextual Environmental Ethics

Contextualism

Contextualism is a broadly applicable term for ethics that places significant weight upon the context or situation in which an action is contemplated or was performed. Just as we live and decide what to do within a context, we must make our moral decisions in terms of what is fitting in that context. The rightness or wrongness of the action is affected by the context, all the aspects of the situation that can have moral relevance.

The context includes many things, historical, cultural, and natural. The moral agent, in making a moral decision, will have contact with many factors within his or her lived world. The natural factors, as they are constituted by the agent who has access to the "real world" only through consciousness of it, are very important in decision making. I do not think an abstract system of morality, which would base moral decisions just on logical, theological, or metaphysical notions, can be adequate. An approach to ethics based on concepts of the human person that are not grounded in the best knowledge available to us about individual persons and human societies should not be trusted. Knowledge of the natural world, based on study of that world itself, is essential. Generally, I believe, the more we know of history, along with knowledge of our culture and other cultures, the better prepared we are to make moral decisions.

There can be disagreement, of course, about which aspects of a situation have moral significance and which have application to a particular moral judgment. It is easy to see that in a case of telling a lie on a Tuesday afternoon, the Tuesday and the time of day are not the relevant matters. Lying at any time is relevant, while the day on which an action occurred is seldom morally relevant.[1] In other contexts the morally important aspects may be hard to distinguish from those that are not morally significant. Trying to explain why this or that is morally relevant may not be easy, and fortunately it is not always necessary. This may be one of the things that often rests on the strong shoulders of good judgment. It also helps to take care in the way we describe things, whether a specific identification of a person, for example. or a more generic reference.[2] We can obscure the moral relevance of an important factor by the way we describe it. I have heard pieces of land described as "disturbed" by developers who wanted to build on it, while other people were trying to preserve it because it was close to being pristine. What to one person is just widening a neighborhood street to ease traffic flow is to another person cutting down majestic old trees. Is it morally relevant that the trees might be cut down? It might be judged necessary to cut the trees, but who can honestly say they are not part of the context in which the decision is to be made? There will be disagreements over what is morally significant in complicated cases and in very new situations, but that is to be expected and is not a completely bad thing, as I will try to explain in Chapter Ten.

There have been a number of versions of contextualistic or situational approaches to morality. Some may not refer to themselves as contextual but speak instead of narrative or history, but these are, I believe, ways of speaking of the importance of what I call context. The several forms of contextual ethics have some aspects in common, but their differences are important. I prefer the term *contextualism* to another term, *situationism*, which is frequently used for ethical approaches that acknowledge that the context of an action has some bearing on moral judgments of actions. Situationism is identified by many people with a particular approach that has been widely discussed in recent years. I do not want the approach I am presenting to be confused with another type of contextual theory, which had some features quite different from what I advocate. Joseph Fletcher's situationism was based on the absolute value of love, which had different moral implications in differing situations.[3] Even though I admire Joseph Fletcher, and I do not want to oppose love, people who know Fletcher's ethics probably identify situationism with him, and his approach and mine are not the same.

The Importance of Contextualism

It is a mistake to minimize the significance of contextualism. One might point out that all ethical systems have paid some attention to the situation in making moral judgments, and to some extent this is the case. In fact, most important ethical systems have recognized the significance of the situation in some way and given it some degree of importance. It is difficult to define the action that is being judged apart from certain aspects of the situation. Saying something untrue to cause a person grievous harm is certainly not the same act as telling a small lie to spare someone embarrassment. Shooting an arrow into a bale of hay and shooting it into the farmer cannot both be adequately described as "just shooting an arrow." The act of setting fire to a wooded area might be good forestry practice and a benefit to the natural environment, or it could be a wicked act of malicious destruction of living things. The context, including the purpose and the procedure, makes all the difference.

The point is that some ethical systems give much more weight to the situation than do others; what makes an approach to moral judgment contextual is that it acknowledges more influence of the situation in shaping ethical norms than most other approaches allow. Contextual matters played a more important role in Aristotle's ethics than they played in Plato's. While Plato talked about justice as harmony between the three parts of the soul, Aristotle dealt with what justice requires of people in various positions in the family and in society. In his description of virtue as a mean between extremes, he was concerned not with arithmetical proportion but with the intermediate "relative to us," which varies from person to person depending on the sort of activity each person is engaged in.[4]

In most ethical systems since Aristotle, context has been recognized as having some bearing on moral judgment. Even systems of ethics based on absolute rules had to reckon with contexts in defining actions covered by the rules. Nothing is clarified, however, by calling all of these ethical systems contextual. In some ethical systems, the acknowledgment of context has been minimal, almost a regrettable necessity. For my contextual approach, the recognition of the context is not minimal and is not grudging. I view the unavoidability of considering the context to be an advantageous matter for ethics. Acknowledging the importance of the context has some significant and beneficial effects on ethics. A contextualistic ethical approach is different in important ways from ethics that does not give much weight to the context.

The best way to talk about contextualism is to look at situations and consider their importance. Let us talk about the decision Lynn must make about selling deceased Aunt Flora's old house to a commercial interest. You hesitate, as you should, to answer so abstract a question. What are we talking about? Is the house of great historical interest? If there is nothing distinctive about the house, historically or architecturally, that simplifies matters. Is the area already developed commercially, or is it a neighborhood of family homes without commercial development? That does make a difference. What business interest wants to buy the property? Will the use of the property be environmentally sound? There are many things we want to know about the situation before we make a judgment about what Lynn should do; that is, most of us want to know more about the context. Some people have a ready answer: any property owner may do whatever he or she wants with the property. No concern about context for these people, but take note of the fact that we are dealing with people whose notion of property rights is contrary to what property law has actually stated for quite some time now, if their notion were ever in keeping with law. We do not need to respect the mistaken notion that property rights are absolute and therefore unaffected by the situation. When I am allowed to raise pigs on my house lot in Boca Raton, I will have to give some credence to an abstract notion of absolute property rights. That is dealing with legal rights, of coursse, not moral right and wrong. What the law allows might not be morally right. What about Lynn? What is the right thing to do? Obviously the context is of critical importance to making a judgment about the morality of selling the property.

The Benefits of Contextualism

One significant effect of contextualism on ethics is to make morals less abstract. If the context is given due consideration, matters of moral significance cannot be approached as abstractions. Contexts are always specific and particular; a context is an actually existing situation in which there are people with their needs and interests and their beliefs and values. Also a part of the context are the possibilities for doing good or harm. Within the context there are resources to satisfy many of these needs and interests, but often there are not enough resources. A context includes the social and economic arrangements that affect people's lives. One aspect of the context we must not ignore is the natural environment, the whole "land community" of which we are a part. In the past we have emphasized the cultural aspects of environment and ignored or minimized the natural environment. Environmental ethics changes that; the new ecological aspects of ethics were developed in response to the growing realization of the importance of the natural environment as part of the context in which moral decisions must be made. Humanistic environmental ethics requires careful attention to both natural and cultural aspects of the context.

The importance of making ethical decisions in terms of context can hardly be given too much emphasis. The way to become clear about moral decisions is to reflect on them with a clear grasp of the situation in which an action would be done. To make a careful survey of all the relevant factors helps us make a well-grounded decision, in contrast to a whimsical, fanciful, or hasty one. When we contemplate an actual situation, or even use our imagination carefully in thinking about a situation we are not experiencing, we are exercising our mental muscles and developing our moral sensibilities. We do well to think about the moral dilemmas about which we read or which we see on television. We develop our ability to make good moral decisions if we think about these things with vivid imagination, which presents us with the kinds of detail likely to be experienced in such situations. It has long been realized that walking in other peoples' shoes improves our understanding. The more we experience of life, the better we can understand it, and an active and sympathetic imagination is the next best thing to actual experience, and in many cases it is the only thing available as well as being the only safe access to some aspects of life.

There have been philosophers who tried to make ethics as abstract as possible. Plato believed that understanding came with contemplation of perfect and unchanging entities that were not a part of the world of our daily experience. It is not hard to see why he found these entities more attractive than the things of the world. Abstractions can be exact, and they can be very clean. Some minds revel in the exactness and certitude of abstract formulations. These philosophers can make significant contributions to some areas of thought, expecially those in which a high ratio of symbolization to ordinary language in a journal article is a mark of success. Ethics does not lend itself very well to this kind of exactness, however, as Aristotle realized long ago. He said that the degree of exactness that is warranted varies with the subject matter. He saw that ethics cannot be an exact science. Ethics deals with the behavior of persons within a situation. There is no way to make this neat and exact.

Contextualism denies that morality places obligations upon us that are not related to the context in which we live. The context determines the significance and the effect of our actions. Abstract statements of moral principle are incomplete apart from the context in which moral agents must be honest, brave, and loving.

An aspect of contextualism that bothers some people is that it does not allow for absolutes in the form of specific rules to be obeyed at all times and places. Even if some general principles are accepted as moral absolutes, the context determines how they are to be carried out. So general a principle as the duty of nonmaleficence is not immune to the influence of context. We need to know the context to know whether pain was inflicted gratuitously. A physician I once knew enjoyed removing bandages in a way that caused the patient considerable pain as well as anxiety. That was gratuitous infliction of pain, maleficence. Another physician of my acquaintance sprayed bandages with a solvent to make them slip off easily. Would a physician be acting maleficently in ripping off bandages in a situation wherein no solvent was available? Can we even begin to know whether someone is preserving or destroying the stability and beauty of a natural system unless we know something about the context? In fact, many of the disagreements about treatment of the environment hinge on factual information about a particular site. Contextualism brings flights of abstraction to ground very quickly.

Of course, contextualism has to recognize as right in one context what was wrong in another. Why do many people object to this? Some people probably find it too much trouble, requiring too much thought. Others cannot get beyond a notion of morality they were given as children, one that might have been sufficient for them as children but is not adequate to meet the demands of adult life. Some people, perhaps most of them philosophers, have an exalted notion of morality, which was upheld by many philosophers in the past; they want morality to be exact, certain, and far neater than life itself. Contextual obligations are seldom certain because most situations involve elements of uncertainty. Other people seem to fear that somebody will get away with something they ought not; they see contextualism as a nest of loopholes.

Contextualism cannot be defended against all these charges. Some of them are nothing to be ashamed of. No apology is needed for morality requiring maturity and the use of investigation and thought. Perhaps those philosophers who fear that contextualism will be the death of morality can be given some reassurance, and I will attempt in Chapters Ten and Eleven to explain that a contextualistic, humanistic holism will be an adequate moral system. Those who believe that contextualism is an easy and very permissive ethical system should not get their hopes up or their fears raised. Within a context ethical requirements can be specific and clear, perhaps more so than in ethical approaches that minimize the importance of context.

The environmental ethics I am defending is contextual. Even more, it recognizes that the context is that of people's lived worlds. How could it be otherwise? Only an extremely abstract approach to ethics that spins fancies about imaginary and fictional worlds can avoid what Edmund Husserl called "returning to the matters themselves." The lived worlds are the only real worlds to which we have access. The time is past when we could talk meaningfully of things as they really are, in distinction to things as we humans constitute them. When our choice is between fancy and fiction on one hand and lived worlds on the other, there seems to be no choice for the moral philosopher but to attend to the content of the lived world and deal with the lives that people are living in the contexts in which those lives are being lived. It is the world view of the moral agent, the person who must decide what is the right thing to do, that shapes the meaning events have for the agent; it is within this context that the moral obligation has significance and is felt to be binding on the agent.

We must be careful, however, not to make this talk of lived worlds sound too individualistic. In moral judgment, as in other aspects of life, intersubjectivity is our safeguard against needless error. Ethics within the context of lived worlds need not be idiosyncratic. Our lived worlds are not individual isolation chambers. The world that comes to us through the constitution of individual consciousness is one world, or at least intersubjectivity indicates that it is one world. The grounds for thinking it is one world are much stronger than any grounds for denying that we are all part of one land community.

The lived worlds of those who are open to other people and to the natural world are not always neat, and they are not directed by absolutes one need not think about. Life is relentlessly daily, and we are often a lot of trouble to ourselves, but this is what we must deal with, as responsibly and wisely as we can. The alternative of accepting an artificial view of life and the world is not an attractive one. There are ways of escaping the trouble of thinking about ethics in the messy world known to us through consciousness; we could surrender to the authority of a myth, a tyrant, or a dogmatic society. People are doing these things all over the world right now. We could do one of these things, but it involves such a basic dishonesty and such a denial of freedom and responsibility that it cannot commend itself to us as a moral mode of coping with life.

Contextualism and the Hard Cases

It is in making decisions about the hard cases that we can see most clearly the advantages of a humanistic and contextualistic approach to a critical holistic environmentalism. In "Animal Liberation and Environmental Ethics: Back Together Again," J. Baird Callicott[5] has demonstrated such an approach to decisions regarding domestic and wild animals. Responsibility for protecting habitats for wild animals and for not causing unnecessary pain to domestic animals can be reconciled, bringing back together again ecological ethics and concern for animal rights. Using Mary Midgley's concept of the mixed community of humans and other animals, Callicott shows how the principles used in relationships with other humans can apply to treatment of those animals who belong, historically and in contemporary practice, to the mixed community. Wild animals are of moral significance, not because we are responsible for their condition to the same extent we are for our pets and livestock but because of their importance to the biosphere. We are not ordinarily responsible for feeding them and taking them to the veterinarian, but we must not cause them gratuitous pain, and we must not destroy the species by our actions. We have this obligation to animals that do not appeal to our aesthetic sense or to our liking for things cute and furry; we are obligated to avoid exterminating species that crawl on their bellies, live in the abysmal depths, or are too small to be noticed. Those creatures we find unattractive, which some people find even disgusting, might be more important ecologically than one of the mammals we enjoy watching.

The difference between the domestic and the wild illuminates the importance of context. The different natures of these animals, the type of commitment we have implicitly made to domestic animals, and ecological considerations are the main moral elements of the contexts. In practice, the distinction between wild and domesticated animals is complicated by the fact that humans have made some wild animals partly dependent on human care. We have extended the mixed human-nonhuman community by what may be seen as a one-sided compromise; we have taken large parts of the animal habitat for our own use, which makes us responsible for the animals that have been forced into inadequate habitats. Other animals have been fed by humans who enjoyed seeing them close at hand, and now these animals are dependent on human feeding. Other animals are seriously affected by our introduction of exotic species into their habitats.

How do we resolve the issue of feeding dependent animals? How do we decide whether to move a herd away from land they are overgrazing? How do we decide whether to take a hand in protecting one species from another? We should expect different answers to the questions from animal rights activists and from environmentalists. The concern of animal rights people to prevent pain might lead to a program the environmentalist will think mistaken. Many of us will be caught in the middle, sharing the concern to avoid pain and the concern to preserve a stable ecosystem. Let me suggest that this is not a situation that calls for hierarchical or binary thinking. We can see quickly that the animal rights advocates and the environmentalists have more in common than this division of thinking indicates. Their practical programs often coincide, as in their shared opposition to the importation of exotic birds. They also have a common opponent in the person who rejects all concern for animals as of no direct economic value to humans, holding that people or business as are the only important interests.

Let us return to a hard case, a case that sets a friend, with whom I carried signs objecting to animal testing of cosmetics, at cross-purposes with me. How does a contextualist ethics handle this? We know that the feral goats are destroying the area in which they have an expanding population; we do not need information. The question must now be faced on normative grounds. As situationalists we can decide together that if the animals are removed from the area, they will not be made to suffer unnecessarily. Letting the local bow-and-arrow brigade shoot them can be ruled out; it will cause avoidable pain and it is not necessary as a means to remove the animals. The environmentalists can attempt, possibly with success, to convince the defenders of animal rights that the environmental concern is urgent; the ecosystem is being destroyed and will soon not even support the goat population. The people concerned with animal suffering, which would include many of the environmentally oriented people, can try to persuade the environmentalists to prevent as much suffering as possible, and they might succeed in this. If no place can be found for the goats, destruction of the animals can be accomplished as humanely as possible. If a compromise can be found, what will be its basis? Several things can play an important role in deciding what to do: the need to move the goats is urgent, painful killing of the goats is not necessary, and spending money (which the environmentalists would like to spend on habitat for wildlife) for a merciful end to the crisis is morally right because it is probably the fault of humans that the situation developed in the first place. We should not assume that disagreements between defenders of animals and defenders of the natural environment can never be resolved.[6] I believe the openness and flexibility needed to make contextualism work will facilitate the resolution of many difficult disagreements.

Other hard cases are involved in the matter of human population control. The need for preventing great increases in human population seems too obvious to deny. From an environmental perspective, an even smaller human population is desirable. Attempts to deny this overlook important matters. Making the claim that we have the knowledge and technology needed to provide food for a much larger population does not reckon with the problems of destruction of forests, wildlife that would be made extinct, and the shortage of drinking water. The problems are real, and the solutions raise complex issues. This is a prime case of the difficulty of reconciling humanistic and environmental concerns. There are serious conflicts between human liberty and the controls that are needed. There are conflicts between socioeconomic groups and national cultures and economies. Meanwhile, population increases exponentially.

How can a contextualistic ethics approach the issue? One of the ethically significant matters involved is human liberty. Simply to reject concern for liberty would not only be antihumanistic, it would be to approach the issues abstractly and reductionistically. Any solution arrived at in this way would be draconian, unfair to large numbers of people, and probably ineffective. How can human population be controlled without losing reproductive freedom? First, it is necessary to see reproductive freedom as something other than an abstraction. What does it consist of? Are we dealing with the positive or negative aspects, freedom to reproduce or freedom not to? Who has reproductive freedom and to what degree? Age, sterility, and health problems prevent many people from having substantial freedom to reproduce. In many societies there are effective social controls over reproduction. How much freedom to reproduce are we trying to defend? Does the desire to have a very large family give a person some sort of right?

When we go beyond seeing rights as abstractions and absolutes, we realize that desires do not translate into rights. Who would grant that a desire for slaves becomes a right to enslave? The desire to have a large family may not create a right. The contextual approach calls for serious discussion of the limits to a right. What an absolutist might approach as an unquestioned right, the contextualist would have to examine in terms of the harm done by the exercise of the claimed right. Some sort of compromise is to be expected.

If the compromise between reproductive freedom and the need to avoid excessive environmental harm calls for one child to a couple, there must be a decision about bringing about acceptance of this limit. Positive means, rewarding those who comply, and negative means, punishing those who do not cooperate, might be employed. Which would be more effective? This is a question for the social psychologist. Which approach would limit liberty less? The positive approach might seem to preserve liberty, but when we see this matter in the context of socioeconomic conditions under which people live, we see there is a strong element of coercion in offering poor people a reward for having few children.

The affluent can afford liberties that the poor cannot afford. A negative approach using penalties might be fairer than a positive program of rewards.

One argument, which is convincing at first and appeals to our humanistic feelings, claims that industrialization of Third World countries will lead to smaller families. When we look at the total context of the population issue, however, the appeal of this benign-looking approach fades. The industrial nations, especially the United States, cause most of the pollution and use most of the natural resources. When impact on the planet is considered, we see that the United States is the most overpopulated country in the world. The planet cannot afford any more industrial nations on the American model; in fact, it may not be able to afford what we are doing much longer. The problem of overpopulation and the problem of overconsuming populations are stressing the planet. A few of the industrial nations have a birth rate that will accomplish a decrease in population, but this is not enough. The need to develop sustainable societies is pressing. We need not plan to woo people away from large families by providing fast cars and ranch houses. Seeing things in context may not provide us with attractive answers quickly arrived at. What it can do is show us that some answers that appear attractive will not work, and this is an important contribution. Seeing things in context will also give us a clearer understanding of the programs that are sufficiently effective and sufficiently just for our consideration of them as we prepare to make some hard choices.

In most of our communities we will need to make decisions about the use of land. There will be conflicts between plans to use the land in an environmentally beneficial way and rights that people claim on the basis of ownership. Some landowners will claim an absolute right to make the best use of their land, which they will usually understand as the use that brings them the most money. What can a contextualist contribute to this conflict? One thing is the historical context. Few people realize that landownership as we know it is a fairly new thing. Throughout most of human history, there was nothing like our private ownership of land. Hebrew culture viewed the land as belonging to God, who assigned it to tribes for their use, but not for real estate transactions. Land could not be permanently alienated. In other cultures the ruler or the church owned the land. Even in modern times in some Western countries, possession of land was based on use, and there was no right to pass land on to one's children. The emerging notion of ownership was first based on claims growing out of use and improvement, as seen in the social philosophy of John Locke, who tried to give a philosophical defense of private ownership in the eighteenth century. As we developed the concept of private property, ownership was restricted in various ways. In most of our towns and cities, the owner of a lot is not allowed the best use of the land, if by that we mean any use by which the owner wishes to make money. Matters of health and public safety restrict the use of land. Seen in historical context, restriction of land use in the interest of environmental preservation does not seem very radical.

A contextual approach to land use will see the issues of private ownership rights and the needs of the whole society, along with the emerging sense of responsibility to the environment, as claims to be reckoned with. This approach will rely on the knowledge of trained people in ecology, law, public health, and ethical principles and methods. Finally, it will rely on calm judgment. Making decisions is not easy, but it will be possible to find a satisfactory compromise, which absolutists seldom can do.

A workable approach to hard cases must recognize the goals and purposes of all people involved in the decision process, take account of all that is known about the situation, consider the consequences of all proposed solutions and creatively search out alternative solutions; then all these matters must be carefully weighed and a decision made on the basis of careful judgment.

Am I claiming a contextual approach makes resolving such situations easy? Am I saying everyone will like the resolution? No, indeed not. The decision might be only the best we can do, but it is a resolution that accomplishes some of the purposes of both groups. The point I want to make is that it is easier for contextualists to face such a dilemma creatively than it is for people who take an absolutist position. People who try to see everything that is morally relevant to the situation, who try to make practical judgments aimed at the best consequences, and who can accept the fact that good people do not always agree might work things out in such hard cases. People who are moved only by sentiment, and never by practical judgment, will not be likely to work out a solution. People who always "stick up for principle" are not likely to find the solution; the in-spite-of-Hell-and-high-water people can seldom find their way through the twists and turns of real life. The absolutist does not respond to a real situation but to an abstraction. There are many faults in an abstract view: it tends to be simple, while life is complicated; it tends to overlook the stubborn parts of a situation; and it fails to recognize the complicated nature of the humans involved in a situation. The last point is very important. Our abstract designation of people according to positions they take on issues must not be taken too seriously. I will never forget seeing a deep ecologist tenderly hold a dead little bird in his hands, nor will I forget the support that animal rights activists have given to land preservation efforts in my town. Rather than being a messy, unsettled, and morally compromising way to approach hard cases, contextual ethics has a touch of Solomonic wisdom the absolutist cannot achieve. Questions such as medical research on animals, the use of animals as organ donors for humans, and the use of animals for food will be divisive for a long time because the ways people approach these issues are based on differing principles. Population and land use issues will require careful and creative thinking and the acceptance of new principles and practices. It is highly unlikely there can be an ethical system that enables people to settle these issues cleanly and definitively. The best ethical system will be the one that offers the most effective way to approach hard cases. We cannot wait for a perfect ethical system. We need to rely on approaches that allow for resolution if not perfect consensus. Part of what a contextual approach includes is recognition of the importance of urgency, and the problems that face us are urgent. Unlike the slash-and-burn agriculturalists of earlier times and more primitive cultures, we know we have this one planet and no other. We must act for people and for the planet. A context-oriented humanistic holism can give the best guidance available in meeting this responsibility.

CHAPTER NINE
Moral Pluralism

Recent Support for Pluralism

Pluralism has received serious attention lately, with several moral philosophers supporting the concept of multiple sources of moral insight. Bernard Williams argues that morality has suffered not from too much diversity in moral guidance but from too little. Eugene C. Hargrove, in comparing nonconstitutive rules in chess to moral rules, suggests the use of rules that cannot be reduced to one general principle. Andrew Brennan cites the inadequacy of utilitarian and deontological theory for environmental ethics and advocates "ethical polymorphism," with a variety of moral principles that fit different situations. Gary E. Varner argues that holism must be pluralistic if it is to judge both individuals and ecosystems morally considerable.[1]

Detailed explanation of how different moral principles, which cannot be reduced to a single unifying principle, can be applied to different ethical situations ranging from relations with family members to one's obligations to the human community, and including duties to future generations, animals, and the biosphere, are given by Christopher D. Stone and Peter S. Wenz.[2]

The pluralism described by Stone and Wenz should not be confused with an approach we have known about for a long time. This tolerant approach accepts very different behaviors as morally justified because their contexts differ. A famous example of this is the acceptance of the killing of the aged in those cultures in which the beliefs about immortality and the rigors of daily life make the killing an act of devotion and respect for the aging relative. The point made by this approach is that the moral principles involved are the same as those accepted generally. The behaviors differ, but the principles are the same. This approach is better understood as situationism or contextualism because it does not involve the use of differing basic moral principles in the different situations.

Another approach that is sometimes called pluralism should not be confused with the pluralism of Stone and Wenz. This type of ethics, which accepts as morally correct differing actions and principles, arises from the acknowledgment of ignorance or from skepticism. Not knowing what is correct, the tolerant moral agent refrains from making a judgment and accepts all positions as right. If this is just a matter of not knowing the best means to an agreed-upon end, it is not a very deep form of pluralism. It can be a denial of the possibility of making moral judgments, and this is not what pluralism does. Certainly Stone and Wenz do not advocate an approach to ethics that accepts blatant contradiction, which would be the result of accepting all moral claims.

The form of pluralism that might be the most effective approach to environmental ethics accepts different basic principles that cannot be made compatible or reduced to one principle. It is not an attempt to show that seemingly contradictory behaviors can be justified by the same moral principle. It is not a form of tolerance growing out of skepticism. Moral pluralism holds that the several moral principles are justified; they are not accepted because we cannot decide between incompatible principles. As they are used, each principle is a true and adequate moral principle, even though each principle is independent of other principles that are also used in different contexts and for different kinds of moral judgment. Each principle is basic and cannot be seen as merely a corollary or variant form of a more basic theory.

There are varieties of moral pluralism, at least two of which should be distinguished before we continue. We can identify a soft pluralism, which accepts the use of multiple moral principles because of our inability to find an underlying principle that would unify our system. Anthony Weston argues that environmental ethics is in an "originary stage," a period in which it is in process, marked by experiment and uncertainty. It is too early for the ethical monism Callicott calls for (see the next section of this chapter).[3] It is important not to confuse this use of multiple principles with moral skepticism. This is not moral skepticism because the several principles employed are believed to be fully justified, principles, but there is the hope that eventually a way can be found to unify the principles. With more analysis they may be seen to be derived from a more basic principle, or they might be discovered to be parts of a complex but cohesive principle. This kind of pluralism would be seen as a response to an epistemological problem. This soft pluralism would probably be adopted with the hope that it is a temporary expedient that allows us to take into account all that is morally relevant while we search for the unifying principle.

The pluralism advocated by some current thinkers seems to be a hard pluralism, which does not consider the use of diverse principles a temporary expedient. The pluralist is willing to accept principles that cannot be reduced to a common foundation. This pluralism is not treated as only a practical necessity, adopted with the hope that eventually a unifying principle will be found. The hard pluralist does not express a desire for such substantive unity. The pluralist need not hold that unity based on one general principle is impossible; there is no need to claim such omniscience and make predictions about what will and will not be discovered as we continue to work on ethical theory. The point is that the pluralist does not see the lack of a unified foundation as a fault. Having one basic principle is not treated as a desirable feature of ethical systems, so the hard pluralist is not a reluctant pluralist who would prefer to have a different system. The hard pluralist sees availability of independent principles as an enrichment of ethics.

The use of pluralistic ethics requires a responsible approach that provides justification for using the principle or principles employed at any time. What is needed is some rationale for using diverse principles and some guidelines to keep the selection of principles from being arbitrary. I think of this as a structural unity in the ethical approach, in contrast to a substantive unity of having only one underlying moral principle.

The hard and soft pluralist would probably not make different moral judgments about an environmental issue, so the difference between them might not be of obvious practical consequence. I think the hard pluralist has the advantage, however, of seeing the ethical approach as a strength and not as an unavoidable weakness.

Some strong support for pluralism comes from ecofeminists. Carol Gilligan's 1982 study of the difference between the way women and men think about moral matters, *In a Different Voice*,[4] drew attention to a source of moral insight that had been neglected because the voices of women did not fit into a monistic approach to ethics, which favored a rule-oriented system founded on rational argument. The concern of women about relationships, caring, and finding alternative ways to solve problems was not treated with much respect or even recognized as belonging to ethics. Karen Warren has advocated a moral pluralism that enlarges ethical dialogue by listening to the voices of people from different backgrounds who have had different moral experiences. She advocates an inclusive ethic she describes as a collage, mosaic, or tapestry of voices, in contrast to a unified morality that controls belief, values, and conduct.[5] This should not be taken as her rejection of the importance of ethical theory, but she favors "theory in process which will change over time."[6]

From the perspective of deconstructionist, postmodernist epistemology, Jim Cheney argues against "totalizing theories" in favor of ecofeminist ethics based on "bioregional narrative" that allows the "multiple voices of this Earth" to be heard, speaking in an "ethical vernacular."[7]

The feminist support of moral pluralism shows a rejection of interest in a common principle underlying all moral judgment. Dependence on one unifying principle is seen as a reason why the voices of oppressed people have not been heard.

Christopher D. Stone (1987) and Peter S. Wenz (1988) have given detailed descriptions of pluralism, with careful explanations of why different principles are appropriate in different situations and with different kinds of moral activity. Stone summarized his position in an article in *Environmental Ethics* (1988).[8]

Stone contrasts the pluralistic approach to the nature of morality with monistic and determinate systems of morality that he sees in Kantianism and utilitarianism, which are monistic in that they are grounded in a "single and complete set of principles" and determinate in that they provide "one right answer for each quandary" (*Earth and Other Ethics*, 116). In his *Environmental Ethics* paper, Stone says that an important consequence of monism is that environmental ethics would employ the same principles on which obligation of people to each other are based (143).

In his book, Stone offers two main reasons why a monistic approach is not adequate and why a pluralistic approach is justified. The first reason for selecting pluralism over monism is that ethics includes several different activities, such as making personal choices; prescribing conduct for others; judging past actions; and judging character, emotions, and attitudes. These activities are so different that the basis for judgment need not be the same for all of them (118ff). The wide diversity of the things toward which we might have moral responsibilities provides the second reason for rejecting monism in favor of pluralism. We are morally responsible for our treatment of living and future generations of people, nonhuman animals, and even Earth itself. The reason for acknowledging moral responsibility for one kind of thing might not be the same as the reason for feeling responsible for another. We might recognize that humans are morally considerable because of human qualities, such as rationality, while we are responsible for nonhuman animals because they are sentient. We can at the same time have moral duties toward wild animal species because of their role in the biosphere. Making a single moral principle account for these varied sorts of moral duty deforms the principle by forcing it into serving where it is not cogent (122f).

Stone suggests that ethics might be more like natural science, which does not attempt to study all phenomena within the same discipline (129). Not taking a pluralistic approach will impoverish ethics.

> The commitment to make every moral judgment responsive to the dictates of one Big Comprehensive Theory forces us to disregard some of the data, to settle for an increasingly bland generality in our rules, and to estrange our moral thought from our considered moral intuitions. Hence, the more we try to force every diverse dilemma into one common mold, the less we can expect detail, insight, and direction in the real, concrete choices we face; the less will our final judgments carry, and deserve, genuine conviction. (130f)

Stone's concern is that ethics not be reduced to one set of basic principles, the way science would be reduced were psychology limited to using the principles of physics. Stone has been criticized for comparing pluralistic ethics to the disciplines in science, as some responsible scientists do try to unify the sciences and moral pluralism might not be as much like science as Stone treats it. This criticism is peripheral, however, because the analogy with science is not crucial to justifying the ethical approach. No one should be disturbed to find that ethics and science are very different. The question is whether ethics is made poorer if forced to apply the same principles to different activities and situations and enriched by using multiple principles.

Stone does not present pluralism as a form of moral subjectivism. The selection of moral principles to use in a given context is not arbitrary. Stone divides morality into different "planes" or intellectual frameworks, which resolve different specific moral problems (133). Stone seeks to explain the reasons for separating the different planes (also called "domains") as well as the ways to work within them and handle conflicts between them. Each plane has two "fundamental elements." First, there is "an ontological commitment" involving making a judgment about the sort of entity being considered, e.g., sentient beings, human persons, species, or nations (133f). Second, there is the act of determining the proper governance to which the particular aspect of the world is subject, and forms of governance will differ in order to deal with what is "morally salient" (134ff). In *Earth and Other Ethics* Stone compares this kind of moral thinking to the making of maps with special overlays to display specific features of an area (139f).

Two general reasons account for different types of governance in differing situations. In different domains we have "differential access to information," and we "stand in different positions" toward people close to us and toward distant and numerous populations of entities. We cannot have the same relationship with nonhuman animals, future generations, and other non-persons that we have with other human persons. There will not be the same expectations of reciprocity, the same possibility of waving rights and claims, and the same acceptance of risk, as well as other morally significant differences (142–147). In the different domains the logic of moral prescriptions will differ, with a different focus ("fine grained" attention to individuals or a broader focus on larger groupings or ecosystems) and different "moods" (the urgent mood of serious obligation or a less severe recognition of what would be better to do or what would be permissible). In the book, Stone says that by recognizing these logical differences, the pluralist can handle contradictions better than the monist (156–163).

Stone does not expect an ethical system to avoid all quandaries.

> . . . (S)ome inconsistencies are tolerable. . . . a system of rules and principles that can give us some satisfaction in some cases should not collapse if in other cases it should prove not only incomplete, but in some cases virtually contradictory. (171)

Stone gives me the impression he does not see the quandaries and inconsistencies as temporary logical problems. He does not seem to see this as primarily a matter of accepting a degree of logical indeterminacy. The reason for accepting in some cases rules and principles that are "virtually contradictory" to those used in other cases is found in the complexity of the world and of our lives.

Stone recognizes several difficulties that a multiple-value system will face (172–179), but he views the "richness" of moral variables as an opportunity, a "potential to construct a flexible community of moral frameworks, each appropriate to its respective domain" (182).

Earth and Other Ethics goes into careful detail about the theory and application of a multiple-value system of morals, especially in Part V.

In the paper in *Environmental Ethics*, Stone clarifies his attitude toward inconsistency in a moral system.

> Pluralism is not so dogmatic—or perhaps one should just say not so "optimistic"—about the prospects of assimilating morals to (slightly idealized conceptions of) arithmetic or geometry. We simply may not be able to devise a single system of morals, operative throughout, that is subject to closure, and in which the laws of noncontradiction and excluded middle are in vigilant command. (150)

This should not be taken as an endorsement of a system in which contradiction is more than an occasional problem. Stone says that if "out-and-out conflicts were in each and every case endemic to pluralistic methodology, the whole system we constructed would have to be abandoned." If such conflicts were exceptional, we would see them as an indication of the system's "indeterminacy" (152f).

Peter Wenz's approach to pluralism is similar to Stone's approach in several respects, even though he does not accept some aspects of Stone's view.[9] In keeping with his belief that traditional ethical theories fail when used as total theories, Wenz advocates "an all embracing—and therefore more flexible—pluralistic theory" that "contains a variety of principles that cannot be reduced to or derived from a single master principle" (310). He says this has the advantage of using the elements of different theories that are attractive in different kinds of situations (311). Wenz does not suggest using whatever principle one might feel like using at the time, which would court inconsistency. He advocates employing "reflective equilibrium," which modifies theories until consistency is found (310, 313). "Reflective equilibrium" is a systematic way of using good judgment in the use of multiple principles in making moral decisions. It is similar to the good judgment necessary in architecture, engineering, and other areas of life (313ff). Wenz gives substance to the notion of good judgment with his "concentric circles perspective," which recognizes ranges of moral obligation that are closer to or further away from the agent, in respect to the strength and number of obligations in any circle. Within the family and among close friends and associates, there are frequent contacts in which personal acts can be consequential. Other circles are more distant because contacts are infrequent and impersonal, with few opportunities to affect the lives of other people directly. Closeness is not a matter of biological kinship or of race (310f, 316).

The notion of far and near concentric circles helps explain why negative rights are more important in relation to distant people, future generations, nonhuman animals, and the biosphere. On the other hand, positive rights play a large role in close relationships (317, 320f, 324–334). The concept of circles of moral responsibility helps explain why our obligation to wild animals, which is usually to leave them alone and not harm them, is different from our obligation to family pets (325–328). Several chapters in the book are used to treat specific moral issues in detail.

Wenz holds that procedural principles, which are not themselves moral principles, determine the kinds of moral issues that belong in the near and the distant circles of responsibility. One does not arbitrarily decide these matters. The main determinants are the ability to affect other people personally, as in the family, and the possibility of recognizing and providing for positive rights, which requires some degree of intimacy. Keeping promises is very important in the family and among one's close associates. The things that are important among intimates are often meaningless in regard to distant people.

Wenz recognizes the inability of pluralistic ethics to provide definitive answers to all moral questions; he does not think ethics should be expected to do this. Ethics should give "unambiguous answers to some questions, . . . hedged or qualified answers to other questions, and merely indicate the kinds of matters to be considered in relation to a third group of questions" (314).

From this detailed reporting of the approaches of Stone and Wenz, we can see that their pluralism is well thought out, with careful explanations of how to use multiple moral principles responsibly. The pluralism described is not arbitrary or subjectivistic. Where differing principles are used, differences in the moral activity or the kind of situation can be cited to justify using a set of principles not employed in other contexts.

We are now able to define more clearly the concept of pluralism and to understand its attraction for some environmental ethicists. In my analysis I am indebted to Christopher D. Stone, who held that different activities call for different principles and that issues of moral considerability cannot all be based on the same principle (*Earth and Other Ethics*, 118–123).

One aspect of pluralism is seen in applying different principles to various moral activities, such as deciding what to do, making moral judgments of behaviors or character, determining social policy and law, and advocating a moral position. Attempts to base our morality in carrying out these several kinds of activity on a common set of principles have not been very rewarding. An interesting example can be found in efforts to overcome the difference between ethics based on duty and ethics based on the development of good character, but only for some duties and virtues it seems. There are duties that do not seem to correlate with a specific virtue, and there are virtues that do not correlate with a specific duty. Carrying out specific duties and cultivating virtues are both important aspects of morality, but they seem to play somewhat different roles and make different contributions to the moral life. We are seldom disturbed when someone decides to act on principles that are not the same as those the person expects other people to act on, providing that this person acts on more demanding principles than others are expected to follow. There are the much admired acts of supererogation, which lead to a very favorable estimate of a person's character, but they are not considered when determining a person's duty.

We should also notice that utilitarian principles do not play a big role in the evaluation of character, but they can play a role in lawmaking, while character issues are peripheral in this activity. Good character is assessed independently of reckoning achievement, while the law is primarily concerned with achieving good consequences, regardless of the character of the citizenry. In relations with family and close friends, a utilitarian judgment will almost certainly be out of place. Having made a promise to one's child, one does not start calculating the utility of keeping the promise. The social policies that regulate the behavior of people who are strangers do not work well in personal and intimate relationships. It is in making social policy that utilitarian principles give us the greatest insight.

Any common elements we might find in the various kinds of moral activities are too formal to direct us in the choice of behaviors. They are not sufficiently substantial to be the unifying factor in a moral system.

Moral pluralism can be found in the different reasons we offer in explaining why different kinds of things are morally considerable. We believe humans have moral standing because they possess characteristics that are not found in the same degree, if at all, in other animals. Humans are aware of their own existence, can suffer in a wide variety of ways, have hopes and expectations, and can communicate with other humans and share a sense of common humanity with them. Humans can have a sense of moral responsibility and direct their lives according to moral beliefs. These human characteristics have been taken by some philosophers as the ground for holding that humans alone have moral standing. This is not a widely held position among environmental ethicists, however, and it is usually granted that we have moral obligations to nonhuman animals. Why? Certainly not because of their conversational ability. For some people, it is because nonhuman animals can suffer. Others hold that all members of the land community should have moral standing. It has been difficult to settle on one set of principles to account for all ascription of moral standing. We might acknowledge we have duties to the most inconspicuous members of the biosphere because of their value to ecosystems. We humans would not fare well if the same principle were applied to us. Even very formal principles, such as nonmaleficence and fairness, cannot explain why environmentalists consider some beings morally considerable. We seem to need a variety of moral principles to account for the moral standing of a wide range of beings.

Recent Attack on Pluralism

J. Baird Callicott has written a strong attack on moral pluralism.[10] Why does he reject pluralism? He recognizes that some common objections to pluralism are not serious problems. Pluralism might on occasion fail to prescribe a definitive command or even give contradictory commands, but this is not a problem unique to pluralism. Callicot points out the conflict between duties that can arise in Kantian ethics. The charge that pluralism allows a "moral promiscuity," with unscrupulous people selecting moral principles to justify their immoral behavior, overlooks the fact that moral systems are made to guide moral people (110f). Callicott's objection to moral pluralism is much deeper than these commonly recognized problems.

Callicott claims that moral pluralism "severs ethical theory from moral philosophy" in that pluralism allows inconsistency between different moral philosophies on which the various moral principles allowed by pluralists are grounded. This, he holds, is playing "metaphysical musical chairs" because moral theories rest on metaphysical foundations whether the foundation is recognized or not. This concern over a metaphysical grounding for ethics is Callicott's main objection to pluralism. Callicott focuses his attack on pluralism on what he sees as an attempt to ground ethics on contradictory metaphysical systems. He places considerable weight on the need for an adequate metaphysics underlying the ethics.

Callicott objects to pluralism, therefore, because he sees it as an ethical approach that does not have a unified theoretical foundation. He holds that an adequate ethical system must have a unified foundation, and this foundation must be based on an adequate world view.

In some of Callicott's essays on environmental ethics (see his *A Defense of the Land Ethic*), he presented a resolution of the conflict between animal rights and deep ecology, which called for approaching different areas of moral concern, such as family duties and duties in the wider society, with different principles. At least superficially, his approach resembles the concentric-circles approach of Peter Wenz and the approach of Christopher Stone. He compares different layers of moral responsibility to the growth rings in a tree trunk. This seems to be pluralism, but Callicott does not consider it pluralism because he thinks his approach unified by an underlying system of thought.

Callicott defends several ethical approaches that use different moral principles in different situations, including his own, against the charge of being pluralistic. His explanation of why these approaches should not be considered pluralistic helps us see what his objections to pluralism are. Seeing his basis for calling the approaches of Stone, Wenz, Hargrove, Cheney, and feminists in general pluralistic, while he holds that Taylor, Rolston, and he himself are not pluralistic, we are able to determine more exactly what he considers pluralistic. The difference Callicott sees lies in the unifying metaphysics on which his, Taylor's, and Rolston's work is grounded. He says that, on the other hand, Stone's talk about different planes and maps overlooks the fact that duties on each plane imply a different metaphysical background.

Callicott does not accuse specific pluralists of consciously basing their ethics on an inadequate metaphysics. He seems to think that they fail to be sufficiently concerned about a metaphysical groundwork, which allows them to base their views on implicit metaphysical sytems that are inadequate or on several conflicting systems at once. He does not consider the world views that underlie Kantianism and utilitarianism adequate for contemporary ethics because they rest on inadequate metaphysics. Utilitarianism bases the moral criterion only on psychological experience; this reflects the dualistic view of Descartes, which separated the true self from the external world. Utilitarianism has an individualistic concept of the person that fails to recognize the interdependence of all living beings, including humans. Kantianism rests on an eighteenth-century view of the human person that makes being rational (the "Enlightenment equivalent of the *imago dei* inhabiting all human beings") the only significant aspect of being human.[11]

For Callicott, that which distinguishes pluralism from monism is not seen on the level of practices and specific rules of conduct. Diversity in practices and general rules does not make a moral approach pluralistic if beneath this diversity there is a unified metaphysical foundation. The people I have identified as pluralists show no interest in finding a metaphysical source of unity to which the obvious diversity can be reduced. Stone and Wenz do make an attempt to give their approaches a structural integrity, which I will talk about later.

Callicott ties his ethics to a moral philosophy he traces from David Hume and Adam Smith, Charles Darwin, and Aldo Leopold, which he claims gives him a unified foundation for a number of diverse moral duties in widely differing situations. On the ground that they have a sound metaphysics, even though theirs differ from his, Callicott exempts some other moral philosophers from the charge of pluralism. Wenz and Stone, ecological feminists, and others are considered pluralists because they do not have a unifying metaphysics underlying their ethics. Those whom Callicott calls pluralists acknowledge being pluralists, but they think they are pluralists because their moral systems accept a multiplicity of principles. We seem to have two notions of what pluralism is, principle-pluralism and metaphysics-pluralism, one notion used by Stone, Wenz and others, and the other used by Callicott.

The disagreement seems to be over the role metaphysics must have in ethics or over the kind of metaphysical system a moral philosopher must have. The pluralists I have read do not argue in favor of an inadequate metaphysics nor do they deny the importance of metaphysics. What they reject seems to be the concept of an overly ambitious metaphysical system, a grand system that claims to explain everything with certainty and simplicity. They seem to be willing to allow some indeterminacy into their world view. They seem to be willing to be uncertain about some things. Is this to be equated with metaphysical inadequacy and a separation of ethics from moral philosophy? I do not think Callicott has shown that Stone and Wenz and other supporters of pluralism have exceeded what can be allowed in responsible metaphysics. Callicott attacks the deconstructionist approach of Jim Cheney, but Cheney's is just one of several approaches, and Cheney's postmodernism does not provide a common approach for all pluralists, or even for all ecological feminists, only some of whom seem to be postmodernists. When feminists ask that caring and attention to relationships play a greater role in ethics, they are not being hostile to a unified metaphysical foundation; they are for an approach to ethics that does not ignore the voices of more than half the people in the world.

Anthony Weston[12] says Callicott gives metaphysics too large a role in his explanation of the forming of moral theories (283f), and I think he is right. I doubt that all philosophers who accept aspects of morality that are usually associated with Kant, such as the importance of a sense of duty, universalizability, and consistency, also adopt Kant's concept of human rationality. The Kantian ideas do not need to rest upon Kant's philosophy. Moral responsibility and consistency are so basic that most moral philosophers would include them in a moral system. Tying these concepts to Enlightenment philosophy is pointless.

Philosophers who make the utilitarian concern for human welfare a part of their ethics may not be basing this on the individualistic metaphysical and psychological views of Bentham and Mill. They may not accept a Cartesian separation of the self from the world. Because most philosophers no longer support psychophysical dualism, they probably do not rest their thinking on a Cartesian view, even unconsciously or implicitly. I accept the commonly held view that pleasure is good and avoidable pain is an evil. I believe pain and pleasure are morally relevant, and I do not find this belief inconsistent with my monistic concept of the human person and my critically holistic philosophy of nature. The concern about pain and pleasure does not need the support of Bentham and Mill any more than it needs the support of Epicurus's metaphysics.

Some moral philosphers say very little about metaphysics. This does not warrant saying they are basing their ideas on an implicit metaphysics, on the grounds that similar moral views have been associated with a certain metaphysics in the past. Only when the only metaphysical perspective capable of supporting a moral view is the one associated with it historically, are we justified in talking about implicit metaphysics.

I have no argument with Callicott's concern about moral philosophers having an adequate world view. My own ethical theory stresses the importance of a person's individual world view. My theory of the justification of moral claims rests heavily on the requirement that world views be adequate. I think, however, that we should not just assume, without further debate, that complete closure is possible or desirable in a contemporary world view. I am suspicious of theoretical ethical constructs that are too neat and simple. Life is complex and frequently confusing; can a theory that does not recognize that confusion and complexity be adequate for morality? I feel sure it cannot. If we ever learn enough to eliminate all the confusion, we might be able to simplify our moral principles, but it is far from assured that a biosphere composed of interrelated complex systems will ever stop challenging our ability to comprehend. Life might be too dynamic for us to be able to achieve a completely integrated knowledge of the world.

We should try to make our moral theories as clear and consistent as we can without sacrificing our intellectual integrity. There are two aspects of this effort to be responsible, however, in which my approach may differ significantly from Callicott's. The well-developed ethical systems of Wenz and Stone, which offer a clear rationale for applying different ethical principles to cases that differ in ethically relevant ways, are coherent in their structure. They do not say simply that this principle can be used here, another principle there. The difference in moral terms between those situations that involve previous commitments, frequent interactions, and many opportunities to affect the life of another person are morally different from situations that involve large numbers of unknown people. Moral choices about nonhuman animals, plants, and the ecosphere itself are different in other ways, as are choices about future generations. The first point of my disagreement with Callicott is that I hold that a pluralistic system that incorporates clear descriptions of the factors that are morally significant in the different kinds of situations and fits the moral principles to the particular sort of situation is structurally coherent, and this might well be all the coherence that morality need require. I believe the burden of proof falls on the critic to show that a moral approach that is coherent in the ways described is not adequate. The second point of disagreement is that I recognize that finding one unifying ethical principle that is derived from a unified system of metaphysics might not be possible, and I believe that morality can and must go ahead without this one grand principle. We should try to hold as coherent an understanding of self and the world as we can muster, but this understanding should reflect the realities of our lives. Our metaphysics should take account of the way we know the world and ourselves. We should begin with experience, not with a theoretical formulation. Our constitution of a lived world will certainly give our personal world views some degree of coherence, but I see no need, nor even any significant advantage, in forcing our grasp of the world into a neat and artificially simple picture. Life is often confusing and troubling; our ethical thought should acknowledge this part of our experience.

If we make our ethics too much simpler than our lives, we might do very inappropriate things to ourselves, to each other, and to Earth. Indeed, has not this happened in the past? Think of the people who were assumed to be witches and were tormented or killed. Personifying evil and blaming people who seemed strange for the little-understood calamities of the time helped people preserve a simple view of life. Think of the people of color who were abused. Think of the people who denied their sexuality or were forced to accept a limited role in life because they were born female. These things happened because people accepted a narrow theoretical formulation that reinforced comfortable biases and made the demands on each person clear and simple. The search for a more coherent understanding of reality than we now have need not lead us into the dogmatic stances that have caused great harm in the past, but we must be aware of the need to remain open and flexible even as we search for more stable foundations. One of the ways to avoid hurtful dogmatism is to stand up to the fear of our own uncertainty and the need to accept complexity as possibility rather than as obstacle in our search for the good life and the sustainable society.

Morality can make life richer, more expansive, more interesting, or it can make life mean, narrow, and dull. Morality that limits life is what moral pluralists are anxious to avoid, through the use of a flexible and comprehensive approach to morality. Is there anyone among us who does not want a life-affirming morality? Moral pluralists believe that the openness of pluralism is the way to make morality more conducive to life. The critics of pluralism fear it will be a harmful approach. Here, it seems, the burden of proof is on all positions.

Justification of Pluralism

I do not believe that Callicott is right in his criticism of Stone's example of the legislator having one criterion for making laws and another for dealing with family and friends. The making of laws is not the same moral activity as making a personal ethical decision, and different considerations might be germane in the very different contexts. It may be quite right, as many people believe, to base most laws on utilitarian considerations because there is a common understanding that laws must maintain social harmony and welfare among strangers. Most laws govern those less personal relationships in which utilitarian considerations are important to the people involved. If utilitarian practices lead to unjust treatment of individual persons, responsible legislators try to balance demands of general welfare and demands of justice, recognizing that both legal principles need to be taken into account. There is no inconsistency in acting one way as a legislator and another as a mother at home, as long as the difference between the two activities is recognized. As the senator's actions are described, she may have acted inconsistently in voting for a toxic waste bill on utilitarian grounds and against wilderness roads because she believed in a land ethic. The land ethic should have affected her vote on the toxic waste bill. Had the morning bill been one governing the distribution of funds for children's hospitals, utilitarian considerations could have been the relevant ones. There was nothing wrong, however, in her making laws on the basis of one set of principles and keeping a promise made to her child on the basis of another principle.

The main concern should be having adequate reasons for using different principles. Few, if any, moral philosophers would be happy with an approach to ethics that accepted multiple principles without some clear rationale for using such and such a principle in a situation of this or that sort. There must be some basis on which the choice of principles is made. If this basis is not to be a more general moral principle, what can it be? What can make our ethics an ethical system and not just an excuse for doing anything whatever?

Callicott looks to a metaphysical system to make ethics unified and justified. Is this the only way to ground our ethics? Were it the only way, we would have a real problem when a suitable metaphysical foundation is not available. Anthony Weston questions the completeness and adequacy of the metaphysics on which Callicott bases his ethics (284), and it seems to me that there are some metaethical problems with basing ethics primarily on the evolutionary advantage to humans of inherited social feelings. The question of concern to me, however, is not whether Callicott or some other person has successfully grounded ethics on a metaphysical system. The question is whether this is necessary. Ethics based on a metaphysical foundation will only be as acceptable as the metaphysics is believable, and the metaphysics that satisfies one person can leave another unconvinced.

Stone and Wenz attempt something quite different from the metaphysical approach. They provide a kind of structural or functional unity for their ethics through an analysis of the sorts of activities involved in making ethical judgments and the situations in which ethical decisions are made. They then match these activities and situations with appropriate moral principles. What Wenz and Stone are doing is different from an attempt to unify morality around one substantive moral principle or around one metaphysical formulation. It is their use of multiple moral principles that cannot be reduced to one substantive principle that makes their approach pluralistic.[13]

This approach of Wenz and Stone requires careful reflection and judgment, with an effort to eliminate any systematic incoherence. This approach is not arbitrary, but it allows for the consideration of everything that is morally relevant and is open to a wealth of moral values. It is not unified through reducing the values and principles involved to one master principle, but it is unified functionally through providing reasons for using the principle that is used in every situation.

The value of this structural unity is that it shows how a number of moral principles can be used and the justification for using them. Let us explore in more detail the justification for using the principles in the contexts in which they are employed. There are several important points to notice.

One of the points made by Stone is that ethics involves a number of different activities. These require the making of different sorts of judgments. Making morally sound laws and social policies is different from making decisions about what one should do as an individual in dealing with family members and close associates. Deciding what one should do is not the same as making a judgment about a person's character. Even other judgments come into play when one is determining appropriate means to an end. Making decisions involving distant persons with whom one has no personal contact, determining one's duty to nonhuman entities, and being responsible toward future generations require other, different kinds of judgment. Making moral judgments is not always doing the same sort of thing. Employing the same moral principle in all these judgments may be forcing it to serve where other principles would be more appropriate. The claim of the pluralist is that it is justifiable to use the principle that is most appropriate to each activity. If one is conscientiously making laws that govern the distribution of goods between people who will be strangers to each other, utilitarian principles will probably be best. Those principles have a long history of working in such contexts. When the laws will affect the natural environment, utilitarian principles alone will probably not be adequate. They take into account only the welfare of individual persons, and when the natural environment is affected, other things must be taken into account. If ethical decisions are to consider all that is morally relevant, a pluralistic approach will facilitate doing this. Some principles are best suited to deal with some matters, while other principles more adequately take account of other matters. When carrying out an activity, such as deciding what to do in a type of situation not experienced before, pluralism gives the freedom to act on the principle that is germane to the specific factors involved in the case.

A second point made by Stone is that the reasons for recognizing that one kind of entity has moral standing are not the same reasons that show us we have moral obligations to beings of another sort. The moral considerability of human beings is not grounded in the same moral values that give us duties to nonhuman animals. The principle, well understood by Immanuel Kant, that we should not treat rational beings as no more than means to our ends does not explain why we owe the family cat a certain kind of treatment. The pluralist need have no trouble explaining why we are obligated to provide veterinary services to the cat but not to the squirrels who come into the yard, entirely aside from the difficulty of catching the squirrels to take them to the animal hospital for their shots. If different principles support the moral standing of different entities, then these several entities should be treated as morally considerable. It would be quite wrong to acknowledge duties to *x* because its moral standing is based on the one general principle we have decided to accept, while we recognize no duties to *y* whose standing rests of a different principle. Giving unity to our morality in this way seems arbitrary indeed. Doing this is not just a theoretical possibility; just look in the introductory textbooks in ethics that reprint journal articles or book chapters that claim nonhuman animals are not due moral consideration because they cannot use language to form their thoughts and express their claims.

A pluralistic approach seems to me the correct one when we face a situation in which a combination of two or more principles relates to morally significant aspects of the situation. An important example of this is the need for the principle of utility and the principle of justice, which was recognized by William K. Frankena, who held that neither utility nor justice alone is adequate. He called for the combined use of the two principles to make "a mixed deontological theory." [14]

Utility by itself will secure the greatest balance of good over bad for the group of people involved, but this could be done by treating some members of the group unjustly. To do the morally correct thing, application of the principle of utility must be governed by justice.

What Frankena advocated is not the use of one principle at one time and another principle at another time, and he did not treat one principle as more basic than the other. He combined the two principles so that each principle augmented and modified the other. The principle of utility would keep the rule of justice from leading to a merely formal kind of justice in which no one benefited. The principle of justice would rule out an action that did the most good overall but was unjust. When two principles work together as one source of moral direction, they must not contradict each other, but they do limit each other. The principle of justice does not lead to a neglect of utility, and the principle of utility does not call for the abandonment of justice. Both justice and utility must be served but in a way in which one does not neglect the other.

Can we really combine two principles, both of which have been treated by some philosophers as independently adequate guides to moral conduct? Which principles can we combine in this way? As pluralists we do not rely on a more basic principle to guide us in this process. We must use good judgment as we see what conduct various principles call for. When we see conduct we find fitting in the context in which it occurs, conduct that is in harmony with our moral insights, we can approve of the principle that supports it. If we see a social problem solved in a way that does as much good as is compatible with justice for all persons involved, we will approve of the solution. If we see that a way has been found to preserve the natural environment and at the same time treat the humans involved with dignity and fairness, we will approve of the program. It is possible that these delightful states of affairs might be grounded in some metaphysical system or in some more general master rule, but I do not see that this is necessary. If we must justify our actions simply on the basis of the moral fittingness of the approach taken, that is enough.

There are situations in which a moral principle leads us to behavior that is clearly fitting. The principle takes into account all that we recognize as morally relevant, and it directs us to actions that realize the moral value which is available in the situation. The pluralist will find this adequate justification for using the principle. It is not necessary to show that this principle will be effective in situations that are very different from this one. The moral monist insists that we must act on principles that provide a foundation for all moral decisions, but the pluralist does not think that complete general applicability is necessary for a moral principle to be justified. The success of the principle in achieving moral value, not its level of generality, is its justification.

Insistence on grounding all moral decisions on a principle of general applicability leads to a search for principles that are so formal they do not provide much guidance in making decisions about conduct. There is no value in finding a principle that can be applied to many different kinds of situations if the price of this general applicability is inability to throw clear light upon actual situations in which we must choose. My students laugh about Marietta's Rule, "Always do the right thing." We do not profit from a general principle that is morally right but that does not give us a sense of direction.

This is not to say that general principles are of no value. They do convey moral insights. They may fail to give us a clear sense of direction, however, especially when we confront matters new to us, such as dealing with entities our parents did not know were morally considerable. We often need to use the revered, old moral insights in tandem with other principles that have closer connections to the actual situations that confront us.

Can we be acting responsibly when we act on principles that are not compatible with each other? We need to be clear about what is involved. What does it mean for two moral principles to be inconsistent with each other? Callicott thinks principles are in conflict when the metaphysical systems in which they are grounded are not compatible. I do not find this as clear and helpful as Callicott does, because I do not think moral principles are forever captive to the metaphysics of early exponents of the principles, as we have seen earlier in the case of moral principles associated with Bentham and Mill or with Kant. It seems to me that moral principles must be seen in relation with the behaviors they prescribe, but these behaviors can be seen as commendable from the perspectives of a number of metaphysical positions. Concern for the relief of suffering, for the dignity of people (and any other living thing that can suffer indignity), for freedom, and for fairness are fitting within a range of beliefs about people and the world. We have seen behaviorists and volunteerists equally concerned about justice and human welfare. Idealists and materialists have been able to join some of the same moral crusades. When Kipling said that East is East and West is West and never the twain shall meet, his cultural awareness was no better than his poetry; on many important issues people from many different cultures recognize the same moral imperatives.

The only significant notion of incompatibility between two moral principles that seems important to me is seen when the principles call for different behaviors in the same sort of circumstance. If the situations in which two different behaviors are called for are not similar in morally relevant respects, the principles are not contradictory just because they prescribe different behaviors. Even one principle could call for different behaviors in differents sorts of situations. This would not make the principle self-contradictory. This sort of thing is expected in a contextualist approach to ethics.

Can a responsible pluralist acknowledge two moral principles that do in fact call for different behaviors in the same situation? Why would both principles be considered morally right? It can be troubling to find that two attractive principles call for different behaviors, both of which seem fitting in the context. It can be troubling, but there are times when being morally realistic and honest might force the recognition that neither principle is wrong. I believe this recognition of moral value in conflicting principles can be a responsible approach to the complexity of life. Some people are quite uncomfortable if they do not have a view of life that allows for complete closure. To have part of the world view incomplete or open is very threatening. These people would probably be uncomfortable with a pluralistic approach. Does that make the person who is able to live with some uncertainty, who is able to live with some openness in the world, morally irresponsible? I do not think it does, because this person may have the most nearly correct view of the world. At least we can see a number of indications that the world is complex and not easily pulled into complete harmony with our orderly pictures of it.

How can two incompatible behaviors both be morally right? Let us see that the problem here is the question, which assumes that there is always a right way and a wrong way. It may be, however, that there is some moral value at stake in both ways of acting. Frequently people see one part of what is involved in a situation. Some people see the current argument over the spotted owl in terms of timber workers' wages. They ask what is wrong with putting food on the table. The issue is complex, however, and the owl is only part of the picture. There may be other endangered species involved, and the old growth forests themselves are important. Loss of income can be traced to management decisions and trade policies, not only to preservation of the forests. The timber yields cannot last long unless harvesting in limited. The forests are important, the endangered species are important, and the people are important. We can save the forests and the forest animals if we try, and we can try to find a sustainable economy for the region. It would be a loss if the area became like desolate regions in many parts of the country that were once thriving communities before a resource ran out. People and their needs are morally significant. If we take all the morally significant matters seriously, we might find a creative approach to the problem.

Callicott thinks that more metaphysical work can unify our morality, but I am less optimistic than he over the possibility of our unifying and simplifying our ethics on a metaphysical foundation. I do not want to argue here about the adequacy of his metaphysics, which I do not think resolves all our moral difficulties, but which I think can make a valuable contribution to our understanding of what we are as human beings. My concern is that I do not think the search for a metaphysical foundation is the most productive thing for moral philosophers to do now. I want to focus on the development of responsible pluralism.

There are several aspects of responsible pluralism. A structural or functional unity, which clearly explains how contextual differences justify the use of different moral principles and explains why certain principles are appropriate in the situations in which they are used, is an important element in being responsible. This structural unification seems to me to be much more useful than a metaphysics that is too abstract to relate specifically to the actual situations we encounter in our living. Metaphysical elements will play a part in our recognition that a moral principle is relevant to a situation, but this might not involve a general and all-encompassing metaphysics. It might not involve one of the traditional metaphysics. What I think will be most important is an adequate personal world view; I write in Chapters Five and Eleven about the adequacy of world views.

Another aspect of responsible pluralism is continued critical examination of the approach to and reflection on the values involved. Because taking into account everything that is morally relevant is important, the system must remain open and the moral philosopher alert to new aspects of value. The moral philosopher must seek intersubjective verification in keeping a personal world view that is enriched by knowledge from all available sources as well as from many avenues of moral insight. The moral philosopher who forgets that our best thought is uncertain and our theories are work in progress will not be a responsible pluralist. Pluralism simply cannot allow dogmatism or complacency.

I think moral pluralism can be a responsible ethical position, and not just as a poor substitute for a monistic approach. Pluralism has several inherent advantages of its own. It is more likely to take into account everything that is morally relevant, including those voices that have often gone unheeded in the past, the voices of women and oppressed people. Where moral monism guards against the adoption of principles that should not be adopted, pluralism can incorporate values likely to be lost under a monistic approach. I believe a pluralistic approach can be adopted because of its moral advantages, while diligent care, critical thought, and careful reflection can guard against errors.

One of the fears of moral philosophers is moral paralysis, which leaves people unable to take any sort of moral stand, and there seems to be some fear that pluralism will leave people confused and uncertain and unable to act decisively. Yet, the notion of moral paralysis might be misleading. Use of the concept of paralysis suggests that movement is not possible. What I see is something different, not inability to move, but difficult choices, even painful choices. It hurts to see that some loss of moral value will occur however one choses to act. Moral action seems to require brave choices in the complex situations that can confront us. In these situations we do not need to be told to find moral unity in a general principle that will make everything fall neatly into place. We need to be told to expect conflict and disagreement. In the face of the disagreement, in the face of the demands of environmental responsibility, we need the courage to choose.

Moral disagreement seems to be inevitable, but it need not leave us without recourse. We do need to learn how to face disagreement, and ways to do that will be the matter for the next chapter.

CHAPTER TEN

Moral Disagreement

Perspective on Moral Disagreement

There are frequently strong disagreements about what should be done about an environmental problem; what one person sees as fitting another person will not approve. Why do we have such disagreements and how should we face them? In this chapter I wish to make several points. First, disagreements are inevitable and sometimes not resolvable, but this is not a bad thing, because disagreement is one price of a precious freedom and it opens up to us the possibility of new insights. Second, disagreements are not all of the same kind nor are they on the same level. In fact, some of them are not primarily ethical disagreements, but disagreements in desire, such as one person's desire to make money and another person's desire to have a nice place to hike. Third, some ethical disagreements can be argued and the right beliefs identified, which is not to say that the correct approach will be accepted by all the people involved. We may expect ethical disagreements regarding the proper way for people to treat the natural environment. When it is thrown up to an environmentalist that many people do not accept protection of the ecosystem as a moral duty, it is important to say that the duty recognized by the environmentalist is not a mere preference, not an optional position for a responsible person. It still strikes me as odd when someone speaks of "the environmentalists," indicating that the speaker is not one of them. I think it will always strike me as strange, irresponsible, and unintelligent. I expect to encounter this viewpoint often, but I see no reason for it to call my position into question. I do not believe rejection of the demands of ecological ethics refutes, or even weakens, the moral claims of environmentalists.

Maurice Mandelbaum, in dealing with the possibility of resolving moral disagreements, points out the difference between resolving disagreements and getting a consensus. He bases his concept of resolution on the generic nature of moral judgment and shows how moral judgments can be evaluated on the basis of moral attitudes and the context of moral claims. Getting everyone to accept the resolution is another matter; consensus may be impossible to achieve for a number of reasons.[1]

Most environmental ethicists have done their research and thinking by tested and productive methods. Their personal world views have developed from open minds and scientific attitudes. They have sought and found intersubjective verification for their factual claims and their sense of value. They understand why there are still areas of disagreement between environmental ethicists, especially on some of the more difficult and technical matters. These philosophers and many other environmentalists who share their moral views need not doubt their ethical positions because there are some people who reject responsible treatment of the natural environment. They do not need to be hesitant about their moral stance or wonder whether their belief about responsibility toward the ecosphere is simply one reasonable opinion among many. Moral disagreement does not justify an extremely subjectivistic interpretation of ethics. Even a contextualist and pluralistic approach to environmental ethics does not lead to the acceptance of all moral positions as correct or as equally justified. Some moral positions are not supportable with evidence and sound arguments, and others can be supported. I do not hesitate to say that a lack of concern for the natural environment is morally irresponsible. This point can be defended from a bio-centric or a purely anthropocentric perspective.

To say that a nonenvironmental position is not morally responsible is not to assert that all environmental projects are scientifically sound or morally right. There are disagreements among those who recognize their responsibility toward the planet; making a decision between one policy and another is not always an easy matter. At times two conflicting views about what should be done can be defended with strong and reasonable argument. At times this is a disagreement about the probable outcome of a program of action, and increased research might settle the matter. A more difficult sort of disagreement occurs when what must be decided is ends and not just means.

Not all environmentalists agree regarding the desirability of certain outcomes of environmental action. For example, environmentalists have not reached agreement about captive breeding to preserve endangered species; some people want to preserve the Florida panther by raising panther kittens in captivity, then releasing them. Other people who care about the survival of the Florida panther do not think this is the best method. The disagreement here may be simply over means to secure an agreed upon end. Consider, however, the concept of preserving the species by breeding Florida panthers with similar species of panthers. Reasonable people can disagree over the end result of such a program.

There is disagreement over the expenditure of scarce funds on the preservation of a species that has little chance of surviving as a species of wild animals in a natural environment. Some environmentalists argue that there are good reasons to emphasize preservation of habitat over the preservation of specific species, but those who place a high premium on preserving genetic diversity have an arguable point. How far should we go in the effort to maintain genetic diversity? Should we preserve a species that will never again live outside of captivity? These are the sorts of disagreements that will arise between people who can respect each other. Scientists can respect other scientists who disagree on these matters. Morally sincere people can have moral respect for those who disagree with them about a specific program. Eventually one of the disputed positions might turn out to be the best one scientifically, especially when the argument is over the means to an end. Other disagreements may be resolved only when the time is past when a particular program is possible, as disagreement over the dusky seaside sparrow is now a thing of the past.

We will probably continue to face disagreements over important environmental issues, sometimes about the means to an end on which we agree, but sometimes over the ends themselves. It is hard to disagree with other morally responsible people, and we will work toward resolution of the disagreements, but there will probably be times when we will continue to disagree on some issues while we work together on those programs about which we agree.

Roots of Moral Disagreement

Why do morally good people have disagreements? There are several ways in which the disagreements can come about. Often a moral disagreement is grounded in factual disagreement. There is much we do not know about ecology, and increased knowledge will put some disagreements to rest. Other disagreements are more complex, because they rest on differing world views; world views, however, are not immune to criticism. World views can be argued; there is a logic to world views, and an ethical position based on an inadequate world view can be exposed. Some disagreements come from basically different beliefs about the nature of ethics; a belief that human welfare is the basic concern of ethics will lead to clashes with people who think justice must be the primary consideration. Thomas Nagel holds that ethical disagreements rest on fundamentally different sources of moral obligation. He identifies "five fundamental types of value," and holds that we cannot simplify matters by subsuming one type under another. Specific obligations to people or institutions, general rights of other people, human welfare, the intrinsic value of certain activities or accomplishments, and commitment to one's own projects cannot be reduced to a common basis. I believe Nagel is correct about this. Fortunately some actions can, in some circumstances, enable us to realize more than one of these values, but in other circumstances we are forced to choose between them.[2]

Nagel also showed that attempts to rank the values in order of preference does not help us much. Even if we could get agreement on putting obligations above human welfare, we would not have resolved all our difficulties. Very consequential cases involving issues of welfare would take precedence over trivial cases of obligation. Having promised to have lunch with one's friend pales into insignificance beside the need to save the life of a neighbor's drowning child. If we established a different ranking, the opportunity to help the neighbor, even a poor invalid, repair the house would not justify failure to keep some promises. However we ranked the sources of moral value, an important case of a lower-ranked value can obligate us more than an unimportant case of a highly ranked value. We will continue to need to make decisions, and we will continue to have disagreements over the relative importance of increasing human welfare, preserving some part of the natural environment, acting in a just manner, realizing the value of education or the arts, and doing something worth doing with our lives. One of our chief moral disagreements is over the nature of morality itself; we need not expect full agreement about the nature of morality to come very soon, if ever.

Differences in environmental philosophy can lead to significant disagreement between environmentalists who are equally committed to preservation of the natural environment. The issue of captive breeding of endangered species provides a clear example of this. Environmental philosophy is not simple; two people who object to excessive human manipulation of nature can have different notions of what is excessive. The people who believe we should preserve natural systems by leaving things alone will not approve some programs of those who believe we have a responsibility to manage what we have damaged. Part of such a disagreement might rest on factual matters, such as the possibility of successful management. Other aspects of the dispute seem to be basic differerences of belief about what is the appropriate place of humans in the system of nature and basic attitudes about the value of unmanaged wilderness. These are disagreements that will remain after all the facts are available.

Uncertainty and Probability

When W. D. Ross, in his book *The Right and the Good*, wrote about choosing between *prima facie* duties to select one's absolute duty, he did not minimize the difficulty; we have no certainty that our choice is the correct one. He said we must "estimate" and reflect on our duties "to the best of our ability," and we are fortunate when we make a right decision.[3] He saw correctly that weighing of moral obligations is a risky matter.

Unfortunately, we have been led to believe that certainty is necessary in morality, but no moral philosophy has been able to provide certainty and definitive obligations in unprecedented situations and hard cases. One effect of this has been to convince many people that morality is just a matter of opinion, that one moral belief is as good as any other, and that view is simply an abandonment of morality.

Even when we are uncertain about which moral judgment is the correct one, or the more correct one, not all judgments are equally attractive to the attentive, insightful, and critical thinker. One action can be identified as probably better than the alternatives in many situations. Fortunately, the alternatives are not always equally supported by good arguments, equally based on intersubjective agreement, and equally in agreement with our moral feelings. We are uncertain about what is morally right, but one moral judgment will have the weight of probability and plausability on its side. In these situations we may not be certain that we are right, but there need be little doubt about what to do. Fortunately we seem to face more decisions like this than situations in which one judgment appears to be as good as another.

Some people have great difficulty with acting without certainty in making moral decisions. Even though acting on the basis of plausability is a common feature of life in serious matters, as when buying costly things such as automobiles, investing our money, and trying to preserve our health, doing something because it is more likely than not to be the morally right thing to do is unsettling. Failure to act on probability is inexcusable in most areas of life. The engineer who does not report a crack in the bridge as a result of uncertainty that it will cause trouble is running a great risk. The physician who does not warn an intelligent patient of probable, serious side effects of a medication is either playing

God or playing the fool. We even want our spouses to let us know when they will probably be late. We will not always know with certainty that a particular danger to the environment will cause the harm that is feared, but we must act quickly to avoid destruction of the reef or the swamp. In spite of the common need to act on probability and the weight of the evidence in most areas of life, many people who have to do what is probably right feel that ethics has failed them. In morality they feel a need for certainty.

Admitting uncertainty, when we use our best knowledge and judgment in making ethical decisions, is not a subjectivistic approach to ethics. If we admit the possibility of error, we cannot have an absolutist ethical system, but neither are we made subjective relativists. The reason the recognition of uncertainty is not the same as subjectivism is that decisions need not be made on the basis of whim or personal advantage. Choices can be made on the basis of the best reasons available. Even though our perception of the facts is grounded in intersubjectivity and is culturally conditioned, our perception is far better than a view that is not based on evidence. Making moral decisions is not like playing poker, in which we win by playing our cards close to our chests and keeping an expressionless face. We engage in moral decision-making best by being open to the collective wisdom of our society and sharing our perplexity with trusted associates. At the level of ordinary moral experience, the level at which we make decisions about what to do, toleration and dialogue and a willingness to learn and grow are desirable. Even though some people with metaphysical or theological goods to sell might say that contextualistic ethics is letting the blind lead the blind, we do not need to accept their assessment. It is a pooling of light and sight so that all can see better. We have no need to wring our hands and lament the lack of universal agreement.

The recognition that a conflict of opinions has not been resolved is unavoidable, but it is not without advantages. Differences of opinion can provoke growth and discovery. It enables us to have an excitingly open and creative approach to ethics. The alternative to an absolute ethics that allows for no disagreement or recognition of uncertainty need not be the sort of subjectivism that requires no real justification of moral principles. We do not need to accept all conflicting actions as equally right.

We must think and act responsibly, with a degree of tolerance that honesty invokes, and then we can argue for our points of view. Occasionally we will win others to our position, and occasionally we will agree with others. Yes, we can change our minds; I have done so and did not find it very painful. I gained the sense that I was on more solid ground than before. Of course, it is easier to change one's mind gracefully and without shame if one has not been too dogmatic at the beginning. I recommend beginning the dialogue with a frank admission that one is not an oracle.

Openness to dialogue and willingness to learn from another should not be an expression of the toleration that comes with indifference; it is amazing how tolerant I can be about things about which I do not care very much, such as the colors of racing cars. This is not a very valuable kind of tolerance. Tolerance can be the responsible position of those who know that we are always in school, learning things we did not even imagine before. It is the openness of those who know that we are always able to grow.

One of the difficult tasks facing us is using judgment about making compromises. As environmentalists we face a situation in which some compromises are responsible and some are not. It is not very difficult to make compromises when nothing more than preferences are involved. He wants a black auto and she wants a red car; nothing is lost when they settle for a bright blue car. In environmental matters, however, we are not dealing just with preferences. It is not that she likes swamps and he likes housing projects. The preservation or loss of a dynamic ecosystem is at stake. We must even be careful about supposed compromises that will allow the destruction of a swamp on the promise of mitigation, whereby another piece of land will be made into a swamp. This compromise has not worked well in most areas in which it has been accepted.

There are several things to keep in mind when contemplating a compromise with some loss of natural areas, possible harm to the water table, serious air pollution, and so on. We must consider the nature and extent of harm that might be caused, whether the harm is permanent or temporary, whether there are means to repair damages, and whether endangered species or rare natural features are likely to be affected. In addition to the protection of the natural environment, we need to consider the effects of the proposed action on people. Often in the past a project that brought substantial money to a few people resulted in long-lasting economic or health losses to a large number of people. Seldom was a serious effort made to make amends to those who were harmed.

One important aspect of compromise is to get beyond the approach of giving each party to the dispute a piece of what is desired. There might be far more creative ways to resolve the dispute. One example of this is sale of development rights. It is not impossible to find win-

win solutions. One thing we must bear in mind, however, is that the human participants are not the only stakeholders. We have no moral right to compromise away the beauty, integrity, and stability of natural systems the way we are free to make compromises when only our own preferences are at stake.

Being Reasonable

Being a tolerant, pluralistic contextualist does not require a wholesale abandonment of the basic moral principles that have guided humanistic ethics for generations. We might have to struggle with the use of them in new situations. We might even have to acknowledge that our understanding of these principles has been limited by our narrow cultural perspective on them. Yet there are some basic principles, such as nonmaleficence, utility, and justice, that will play an important role in our moral thinking. We might not be able to work out some of the problems of environmental ethics on the basis of these old moral principles alone; I am not calling for extensionism. We will probably find, however, that they augment the newer insights of ecological ethics. This is the advantage we might find in pluralism. As we seek to live responsibly in respect to the natural environment, there will be times of uncertainty and there will be some risk, but we will not need to abandon rational decision processes. We may need to go beyond the notion of rationality with which we began, but simply to abandon it would be, I believe, to abandon the enterprise of morality. As we admit uncertainty and make use of those principles that offer guidance in the specific situations which we face, we will not need to be arbitrary or capricious in deciding what to do. We will be able to follow rational decision procedures that are much like the procedures we follow in other areas of life. We can make adequate and defensible decisions. This will not be in spite of the uncertainty and the contextual nature of our judgments, but partly because of them. Uncertainty forces from us our best thinking, thinking that is most responsive to all aspects of the situations in which we live. Having to deal with a specific context keeps us from retreating into abstractions. Facing new situations and the hard cases will be a challenge, but it need not be one to fear.

Uncertainty and the possibility of error need not keep us from well-thought-out decisions and effective action; we have strategies for decision-making even in the more confusing situations. We have developed some widely agreed-upon principles to guide our thinking about the hard cases. We feel obligated as persons and as societies to respect individual rights and to keep commitments, but these principles must give way at times in the face of urgent social needs or clear and present dangers to society. Now we realize there are times when individual rights must be overridden to prevent serious and irreversible environmental damage.

There are no specific rules for determining how great the social or environmental benefit or danger must be to justify overriding a person's rights, but we already know how to get along without specific and exceptionless rules of procedure. We have some time-tested practical rules of thumb that offer us tentative guidance in determining priorities. We find these rules useful because they reflect the common experience of many generations, but we feel free to reject them when we face situations in which they do not seem to be appropriate. We treat rules of thumb as guides only so long as they seem to be working in a given situation, but when there are indications that we should no longer follow these guides, the rational thing to do is to abandon them and think through the situation anew. These practical guides to decision-making are not written in stone, but they are very useful to us in most circumstances, and the fact that they cannot be used in all situations is no reason to ignore their value in the situations in which they are useful.

Our rules of thumb are not themselves moral principles. They are procedural guidelines that help us decide which moral consideration we should act on at a given moment.

These practical guidelines are not new; they are not quickly devised ad hoc rules made to order for pluralism. You will recognize them and remember using some of them.

One rule of thumb for determining the right course of action is based on the level of seriousness in preserving life or avoiding great pain or excessive inconvenience. Saving life takes precedence over the avoidance of pain, but pain gets more consideration than inconvenience. Serious damage to the natural environment must be treated more seriously than inconvenience; such damage can result in health problems, something that can be recognized even by people who are not sensitive to loss of environmental resources. Using this rule of thumb requires the exercise of judgment because the level of loss, pain, or the excessiveness of inconvenience must be judged. Most of us would rather dress and go to the market in preference to the pain of going without dinner and breakfast, but we accept the minor irritation of a scratchy collar rather than drive home to change clothes. In most cases we would expect a person to ignore a minor pain rather than inconvenience a lot of people. These are familiar matters. Environmental loss is a new thing to most people and needs to be made part of our rule of thumb.

In addition to pain and inconvenience, we also need to take account of dangers and possible side effects. No matter how convenient it is to cross the pasture where the bull is kept, we do well to take the inconvenient, long way around. We need to become more alert than we have been to possible side effects in making decisions about the natural environment. We have been slow to learn the danger of quick fixes, which cause more serious problems later. Risk assessment is not always easy and is often a job for professionals, but our rule of thumb warns us to go carefully when we might be taking a large risk. Many people in Florida wish the Army Corps of Engineers had paid more attention to this rule of thumb when they contemplate what the Corps did to the rivers and swamps several decades ago. Now we are more likely to be guided by this rule when we consider beach erosion measures, the location of bridges, and the placement of waste management facilities.

Another rule of thumb takes into consideration the urgency of doing something. Many actions have a period of opportunity, with some actions being urgent, now-or-never situations and others allowing a more comfortable period of time. Of course, life-threatening situations can be very urgent. Other actions can be postponed for a long time without any harm. When my neighbor's little child is in the street, I must act fast. Taking the book to the library can wait. Fortunately most voters in some communities have realized that buying wilderness lands for preservation must be done soon. Doing what is necessary to preserve the Everglades is long overdue; a few more years of neglect and this thriving ecosystem might become greatly impoverished biologically. We are not just joking when we mention the possiblity of the coming great Florida desert.

An obvious rule of thumb is to do the action that has most promise of success. We prefer to help people who are able to derive benefit from our action, and we prefer to help more people rather than a few. After Hurricane Andrew hit Dade County, Florida, many divers rushed to put toppled coral reefs back into place. This needed to be done soon, and the marine biologists said it would be effective. The rule indicated that this was the thing to do. The expectation of success in an effort to preserve a wilderness area makes it the right action to take.

The possibility of remediation and compensation helps us decide between competing obligations. Some losses cannot be recovered; some damage cannot be undone. It is more important to prevent permanent damage than temporary damage. It may be necessary at times to do what we know will cause damage or loss because this is better than causing loss that cannot be made up. We are learning, however, that we need to be very careful about promises to mitigate environmental damage. Developers have been allowed to destroy wetlands on their promise of developing an even larger area of wetlands. Government officials have been more sanquine about doing this than most biologists have been. Recent news accounts report that the wetlands built to mitigate damage to natural wetlands have been more successful than most naturalists had expected, but this good news must be read very carefully. Some sixty percent of the inspected projects are showing promise of being successful if they are carefully watched for the next decade. The sad situation is that although many promises have been made, often no effort has been made to follow up on them, and long-term attention to a project has been uncommon. Simple honesty requires that if we use the rule of thumb about remediation, we must be realistic and vigilant.

A newly grasped factor in applying our rules is the realization that nonhuman animals' deaths and pain must be taken into consideration. There has been some disagreement about the relative importance of pain against fulfilling a role in an ecosystem. It is also a problem to decide between the interests of members of different species, as when the prevention of human pain leads to research that results in death for nonhuman animals. Most people would give priority to preventing human pain, but they might not believe that the death of nonhuman animals is justified for the sake of preventing minor pain or inconvenience to humans.

Do we have any rules of thumb regarding the interests of humans and nonhuman animals? Unfortunately, we have had a rule that we need to question. We have considered nonhuman animals expendable for the sake of any human interest, and now we know that this was not proper. There are some thoughtful and significant arguments for giving matters of human concern precedence over obligations to nonhumans, but there must be limits to this. It is not an adequate rule for governing all situations. Consideration must be given to the levels of interest of a person or a nonhuman; crucial interests must be distinquished from peripheral interests.[4] Killing a nonhuman for human entertainment is wrong because the nonhuman's interest in staying alive is more critical than the human's interest in amusement. With an equal level of interest, as when a human is kept alive rather than a nonhuman, most people will approve of saving the human life. Donald VanDeVeer's treatment of the issue of justice between members of different species provides a thoughtful explanation of giving priority to the human life. In addition to the difference in levels of interest, VanDeVeer adds another factor, the level of psychological complexity. The greater psychological complexity of the human, which makes the human vulnerable to kinds of suffering probably not known among nonhuman animals, helps justify saving the life of the human rather than the life of the nonhuman animal.[5]

One rule of thumb may be used to guide the application of another rule; these rules of thumb are not entirely independent of each other. We have to use good judgment in applying the rules. The rule that tells us to do the thing that will prevent pain is not absolute; we may feel better about inflicting pain if we make it up to the person richly enough, or if some other action is so urgent that we have to let the person hurt for a while.

Our rules of thumb have been used so commonly that we can employ them without giving them much thought. In talking theoretically about ethics, we might forget we have these rules, thereby putting our discussion of ethical problems a bit out of focus. I believe this happens sometimes when people worry about uncertainty in the justification of our basic moral principles.

Do our rules of thumb always give us a clear direction so that we know exactly what we should do? Unfortunately, they do not always give us this much guidance. Especially in what we know as hard cases, we will have to act without precedents to guide us. The moral rules on which we have depended do not seem to fit the novel situation. In confronting these cases in which a great deal is at stake and no action is obviously the correct one, what do we do? We examine the practical implications of our familiar deontological and utilitarian principles and the principles of holistic ecological ethics, and we seek as much help as we can get from our procedural rules of thumb. Then we make a decison as best we can. Of course, this is not something for us to do alone. Eventually everyone is responsible for his or her decisions, but the process of reaching a good decision calls for sharing insights and information.

Facing the moral issues of a troubled world, a stressed environment, and possibly a few personal problems to boot might not sound like much fun, especially with a moral philosopher telling us what we were taught in childhood is not adequate. But we need to approach life honestly. The first aid I learned as a Boy Scout did not include cardiopulmonary resuscitation, the form of life-saving that is now known to be more effective than the old method of pushing the bad air out and letting the good air come in. Would I be justified in using the old method just because I have known it for a long time? No, indeed! It was my responsibility to learn a new method. Why should morality be less demanding than first aid? We must learn what we can and be prepared to make decisions. We are free moral agents, and we cannot escape the responsibility that goes with our freedom. We might sometimes envy the people who lived in more settled times, times in which well-established principles could provide guidance in every contingency. We should not spend too much time envying these people, however, because we have the excitement of knowing that what we do will make an important difference in the world. It is in the context of environmental concern that many of us first realized the critical importance of accepting moral responsibility. Is this responsibility too hard for us? I have not made light of the difficulties, but I do not want to make it sound too grim.

Doing the best we can in making good decisions in important new areas of life is not grim. It is exciting and full of promise. We will make some mistakes, to be sure, but we have quite enough to see us through. We have each other, a resource never to be minimized. We have some fine, basic moral principles that will serve us well if we use them correctly. I believe that means using all of them, in each context using the one that seems appropriate, rather than trying to make one principle serve in all contexts and for all our ethical activities. We also have some very helpful procedural rules of thumb to help us keep things in perspective and to settle priorities. Above all, we have the ability to use good judgment. We have what we need to live in the face of moral disagreement and make good and strong decisions.

CHAPTER ELEVEN

The Moral Adequacy of Humanistic Holism

Moral Obligation Without Certainty

Is a pluralistic, contextualistic system of ethics morally adequate? Does uncertainty keep an ethical approach from being adequate? What is necessary for a moral system to be adequate? These are serious questions, because traditional ethical systems have not been pluralistic and have attended to context in a more limited way than the humanistic holism I am advocating. Some philosophers still argue that ethics cannot function without certainty about basic moral principles and their application to specific situations. Some hold that these moral principles must be held to be true without reference to specific contexts; they must be absolute. I am certainly not alone in supporting a strong version of contextualism, but I cannot claim that it is supported by a majority of moral philosophers. The part of the context that some moral philosophers find especially unacceptable is the knowledge and the beliefs of moral agents. They want moral principles to hold no matter what anyone thinks or how anyone perceives the world.[1]

It is widely held by moral philosophers that moral obligations must be binding on the moral agent; they must be more than just permissive guidelines. I have no argument with that point, but I disagree with the claim that moral rules must be binding categorically, placing exactly the same requirements on everyone, with the understanding that these rules must be independent of the person's beliefs, knowledge, and other aspects of the persons world view. I will argue that moral obligations are binding on the moral agent largely because of the agent's world view. The world view requires these obligations, and this makes them very binding indeed, because the moral obligations are a significant part of the lived world of the person who accepts them, and this lived world of the person is the only world which this person knows.

What seems to frighten some moral philosophers about this is their concern that it will not provide a firm foundation for morals. They think people can change their world views quickly and easily accommodate their adopting an easy or self-serving morality. I do not think, as I will explain, that world views are assumed or changed whimsically. If we are talking about serious and thoughtful people, world views come out of the life of the person and are intersubjectively verified. To be sure, I am talking about serious and thoughtful people who want to live moral lives, and I see no way to develop philosophical ethics that cannot be misunderstood or abused by those who are not willing to accept both intellectual and moral responsibility.

I do not claim the obligations of humanistic and holistic ethics can be categorical in the sense used by some moral philosophers. I claim they do not need to be categorical in this sense to be morally adequate.

The widely recognized requirement that an ethical obligation be binding and universalizable might be interpreted as requiring that it be categorical. Two decades ago, when Philippa Foot objected to this supposed necessity of ethics requiring categorical obligations and suggested instead an ethics of hypothetical imperatives, William K. Frankena attacked her proposal as an attack on morality.[2] Frankena was far from alone in rejecting Foot's approach to ethics, and the concern over the possibility of weakening morality is understandable. Clearly we do not want to make morality a matter of unsupported personal opinion. Morality should make demands on the moral agent. I am not about to support a morality of doing what one likes or a "feel good" morality. That, I admit, would be no morality at all.

On the other hand, I think we need to face the fact that no one has succeeded in making the case that knowledge of moral duty is certain, and I do not think anyone ever will. Of course, some moral philosophers think Thomas or Kant or Mill or perhaps even Aristotle laid the foundation for an absolute morality, grounded in indubitable principles, that can be applied to all our moral problems and give us clear guidance. Even they usually acknowledge there are some problems still to be ironed out in the preferred system. They are much more optimistic than I over the prospect of working out the difficulties in their systems, and I find their arguments for the moral philosophy in question unconvincing. It is not only I who is a doubter. Actually I am not the most severe of the critics or the most thorough of the doubters. So great is the failure of the supporters of any system of absolute morality to convince moral philosophers of the claims made by the sytem that it seems just to say that each of the systems is doubtful. I believe that much of value, indeed some very important insights, can be found in Aristotle, Thomas, Kant,

Mill, and others. But still I think these systems claim too much, especially in presenting themselves not as sources of valuable moral insight, which they are, but as final and complete systems of morality, which they are not.

Let me go further. Not only has no system we have seen suceeded in laying the foundation for an absolute morality with principles that are certain and effective guides to behavior in every situation that might face us, but the very goal of creating such a moral system is mistaken. Let me be so bold as to say that the goal of making morality certain is detrimental to morality. We see an interesting parallel to this in the way some people think of science, especially in holding an overly strict notion of proof. This can be used to get around the findings of science, as when a tobacco company declares that the harmfulness of cigarette smoking has not been proven. A mining company can claim that there is no proof that what it dumps into a stream is harming the fish downstream. What happens is that some people are promoting a false notion of what science is and taking personal advantage of that false notion. Scientific proof is often a statistical matter, which is adequate for many situations. Scientific proof is seldom a matter of "smoking gun" certainty.

One consequence of insisting on categorical obligations is to create a false and impossible notion of what morality ought to be, and this has led to unnecessary moral skepticism. When every moral theory is subjected to criticism from moralists because the theory does not give absolute certainty to moral knowledge, a frequent response is to take moral theories as mere opinion. Add to this the common notion that one opinion is as good as another if no opinion can be proved the true one, and a vicious subjectivism invades the concept of moral philosophy that many people hold. Unwittingly, moral philosophers have collectively weakened moral philosophy. In other uncertain areas of life, this has not happened. Doctors disagree, as we all know, but whom do we call when we are ill? Engineers are not infallible, but they design things on which we depend to go places and to do our daily tasks. We know economists disagree and make mistakes, but most people do better to seek their advice about investments than to trust a hunch. We have not developed as clear a notion of what an expert in ethics must know and be able to do as we have of an expert engineer or physiologist. The concept of the expert in ethics may always involve some difference of opinion. I have no doubt, however, that I and my fellow professors of the subject know more about ethics than the majority of people, who have not studied it technically. Yet how often does someone think to ask a moral philosopher about morality? It is good to see that some hospitals and some businesses have realized the need to have the advice of an ethicist, but this is still rather rare. A number of times I have been invited, sometimes I think as an afterthought, to be on a panel discussing an ethical issue. Are other philosophers on the panel? Not very often, and then only from within the university. Usually it is the familiar priest, minister, or rabbi. What a delight it is to discover that one of these has had a real course in ethics! It is also a delight to see more than a handful of people attending the meeting. I fear that much of the public knows that the ministers of religion usually will have their own brand of morality to offer, and they have no idea what a philosopher can contribute, so they stay away and feel justified in doing what they wish because it is all a matter of opinion anyway. We might one day learn to help people more if we approach ethics with more humble claims.

Meeting Requirements for Moral Adequacy

It is not enough, however, for me to argue that the quest for certainty cannot succeed and that the attempt has harmful consequences. It is not enough to point to the failure of attempts to justify ethical absolutism. It could be the case that a rational approach to ethics is not possible at all. What must be shown is that a contextualistic and pluralistic ethical approach based on fittingness between a person's adequate world view and the person's actions can be an adequate ethical approach.

There are some criteria for an ethical system with which most moral philosophers agree. Ethics is expected to give moral demands that are binding on the agent. The demands must not be individual tastes or self-serving desires of the agent; they must stand over a person, confront the person, and be able to override individual desires and interests. These moral demands must be universalizable, which means that they are not simply individual standards of behavior but can be willed to be universal in application. Moral philosophers have talked about these criteria in a number of ways, but there seems to be wide basic agreement about them. There are disagreements about the interpretation of such concepts as universalizability, but most philosophers accept some version of it. Some moral philosophers have not included among the necessary conditions of an adequate moral system the same degree of concern for human well-being that I would include. I also consider the pursuit of justice to be necessary for an adequate morality. In our time it should be realized that an ethical approach that does not recognize duties to the natural environment is not adequate. I have adopted a group of necessary conditions that includes familiar deontological criteria, such as universalizability, demandingness, and the need to be just. Also included are familiar utilitarian concerns for human welfare. Impartiality, in the sense of not being self-serving, is almost universally recognized by philosophers as a necessary condition; few philosophers support ethical egoism. The only new condition is concern for the natural environment. These are necessary conditions. Some of them might not be germane to a particular moral issue, but each of them must be included when germane. Concern for human welfare, in the broad sense of the principle of utility, might not be involved in some intimate situation, as when a person keeps a promise made to a family member, that does not significantly involve anyone else. In the cases raised by environmental concerns, it seems unlikely that any of the conditions would not be relevant.

Many moral philosophers have limited their list of necessary conditions for morality to the deontological or to the teleological concerns. Some have had some combination of deontological and teleological concerns. It might appear that adopting these widely supported elements of ethics in one ethical system would create a moral minestrone of all available ingredients, but just as a good soup does not call for just anything available, this list of necessary conditions is not utterly without discrimination. Even the moral pluralist can be selective.

Being categorical and absolute are not accepted as criteria of ethical adequacy. While the conditions that are listed are very generally accepted, providing categorical and absolute rules does not have such a wide acceptance. These requirements are not rejected because they are unpopular, however, but because they are not necessary to an adequate approach to ethics. They are not even helpful in a system of ethics, and I will present arguments to show they should not be included among necessary conditions of moral adequacy.

The necessary criteria together provide sufficient criteria for the adequacy of an approach to morality. I realize some moral philosophers advocate a list of conditions they consider stricter than this list of criteria. They also cite fewer criteria. Some of these philosophers think ethics must be able to produce rules that are absolute, categorical, and definitive, giving one answer to any question. Only such a rule seems to meet their standard for a binding moral obligation. What exactly is the proper criterion regarding the bindingness of an obligation that an ethical system must meet to be adequate? Can the obligations of contextual and pluralistic ethics be sufficiently binding?

Obligations that can be ignored if a person does not like them or finds the results of obeying inconvenient are not morally adequate. Moral rules must not be overridden by club rules or social customs. Moral obligations cannot shift with changes in one's likes and dislikes, one's whims and fancies. Some philosophers, however, say moral obligation must not be affected by a change in one's ideals and opinions.[3]

There is a problem with the notion of a person conveniently shifting opinions and ideals. Some people seem to think that people shift their beliefs and ideals very quickly and easily, but these aspects of a world view are not subject to the easy manipulation that some people fear. Of course, some desires are quite transient, but they are not the important part of the world view in the formation of moral beliefs. I have already agreed that individual desires, especially those not shared by the society at large, are not relevant to moral judgment. Should I get a yen for a yellow convertible, this must not be allowed to change my views on automobile theft. The developer of strip shopping centers does not suddenly have a different opinion of the scrubland at the highway intersection.

Significant desires of the sort that can be in some sense universalized can be changed, but this does not happen just because a person wants it to happen. Something usually happens to the person first. A person with no desire to see a section of woodland protected might develop that desire when devastation of a similar piece of land is witnessed. Opinions can be changed, but hardly without cause. If an opinion changes because of the acquisition of some new knowledge, then the second opinion is probably better than the first. When opinions change for the worse, as when a person bitten by a snake decides that all swamps should be drained, the mechanism behind this is easily understood, and we discount the person's opinion. We see it as a flaw in the person's world view. Ideals are not adopted apart from other aspects of one's world view. An ideal is not the sort of thing that one changes as a matter of convenience.

Human personality and the human situation cannot be kept out of moral considerations. The individual moral agent is a crucial part of the determination of moral obligation.[4] It is the adequate world view that enables the moral agent to recognize the rightness of certain general moral principles and specific moral obligations. With a poor world view, the agent may very well not recognize these principles. Acknowledgment of rights and obligations connected with freedom and well-being comes with a world view in which persons are thought of as beings with worth and high potentiality. With a world view that denigrates certain groups of people, there will be no thought of their freedom and well-being. With a sufficiently low opionion of humanity, one might not even value his or her own freedom. A sense of obligation to the biosphere is to be expected with some world views but not with others. Some philosophers cannot accept contextual obligations that are grounded in a person's world view as morally adequate, while I think such obligations are not arbitrary, not individualistic, and not easily avoided. The contextual obligations of ecological ethics meet tests of not being self-serving and not being matters of inclination. These obligations are not disqualified as genuine moral demands because they are relative to the agent's beliefs and ideals, which are among the constituents of a personal world view.

If by a categorical obligation one means a moral duty that is absolute in the sense that it is not dependent in any way on the knowledge and beliefs of moral agents, then the demands of contextual ecological ethics are not categorical; in fact it is hard to think of any demands the moral agent has recognized through moral reflection that would be categorical. There is no reason to insist that adequate moral obligations must be categorical in this way.

Beyond Categorical and Hypothetical Obligations

It is less confusing, I believe, to talk about the nature of moral obligation in terms other than their being categorical or hypothetical. There are better ways to understand the demandingness of moral obligations. It does not get us very far to say that a moral obligation is contingent; what makes a difference is what the obligation is contingent on. It is easy to see that a demand that is contingent on the agent's desires and private interests is not a moral demand; the demand is for a desired state of affairs, but the agent is not morally obligated to secure what is desired. Let us consider, however, a different sort of contingency; the state of affairs called for is contingent on the agent's beliefs for its moral significance. Is this not a different matter entirely? Consider that in this latter case the state of affairs might not be one the agent desires, only one the agent deems morally necessary. Consider a farmer who has come to believe that using large amounts of chemical fertilizer and pesticides is destroying the land his family has farmed for generations. He has heard of some farmers who do not use these chemicals, but his friends laugh at his talk of farming more naturally, more like his grandfather did, and he himself is afraid of what might happen if he tries methods he does not know well. He likes the heavy machinery, likes to drive the big tractors and harvesters. He does not, in a very real sense, want to change his farming methods. He is going to do it, however, because he knows he ought to save the family land. That show on public television made it clear to him. Is his sense of obligation categorical? Is it hypothetical? Neither of these terms seems to say much about his obligation. His sense of duty rests on his beliefs about chemicals used in farming and what people owe their families and the land. These are beliefs he shares with few people in his community. Does this context make his duty hypothetical? I do not think it contributes anything to our understanding to describe it this way. The important thing to see is that his duty is contextual, and the farmer would think of some of the basic principles involved as principles that should be accepted by anyone for whom they are relevant. To save the land is a moral imperative for him and for anyone in his position. What about the farmers who do not share his beliefs? He probably thinks they should share his beliefs and that their view of the land is uninformed. He would be glad to see them act as he does, unless their situation is significantly different from his. Does he consider them immoral? Probably not, if they are doing what they think is right. Can they be clearly wrong in one sense and not wrong in another? This seems to be a good description of the situation.

What does this show us about ethical judgments. We see that the farmer's obligations are not absolute, in the sense of not being relative to a context. They are not categorical, if by that one means they would be morally binding on every rational being, without regard to that person's world view; at the same time, we see that no insight comes with calling these obligation hypothetical. What we will find is that moral judgments are universalizable in a significant sense, which may be the only sense that is useful. I see no value in a notion of universalizability that does not recognize all relevant features of a context.

How would our farmer who was conscientious about the land be judged by Immanuel Kant? Would Kant consider his morality adequate? It seems the part of the farmer's ethics Kant would not accept would be its dependence on a world view. In the Second Section of *Foundations of the Metaphysics of Morals*, Kant stressed the a priori aspect of morals. He said the concept of duty, the ideal of moral perfection, and the supreme principles of morality are a priori. He firmly rejected the notion that basic moral ideas come from experience; what role he left for experience is still a matter of debate. It seems clear that he would not accept a very important role for an individual world view. A world view is largely a product of experience, not a priori; this aspect of the farmer's ethics would probably keep the moral demands that he felt from being categorical for Kant. It is at this point, I believe, that Kant is not very helpful.

The obligations our farmer felt were certainly not matters of mere prudence with him. In what sense were these obligations hypothetical? The were hardly hypothetical for the farmer, if by that we mean they were obligations he should accept if he wanted such and such. These obligations were binding on him because he believed such and such; his world view incorporated certain realities. For the farmer it was not a question of technical or pragmatic imperatives, such as Kant identified as hypothetical.

Because of this confusion over the terms *categorical* and *hypothetical*, I suggest we drop them as we try to discover what is necessary for a position to be an ethical position. These old terms tend to confuse the issues we need to resolve. *Categorical* is used to designate elements that include some that I agree must belong to any adequate ethics, but it also suggests a kind of absolute ethics, ruling out any contextual element. I do not think this absoluteness is necessary for moral adequacy. Kant's use of the term *categorical* inevitably influences our use of the term, and Kant used *categorical* to cover some concepts of morality that I believe to be essential and some concepts that are too limiting. When Kant's use of *categorical* is contrasted with his use of *hypothetical*, it would not seem that a categorical obligation needs to be understood as absolute and not contextual. When Kant spoke of categorical obligations, in contrast to hypothetical obli-gation, he was contrasting them to obligations that are practical necessities if one is to achieve something one wants. He seemed to recognize as categorical an obligation to achieve something good in itself, rather than as a means to something desired. What he did not consider moral obligations were prudential necessities, "technical" and "pragmatic" imperatives. Kant also argued, however, that basic aspects of moral obligation are known a priori, and this gives *categorical* the connotation of being absolute.

As I understand the obligations of ecological ethics. they are not categorical in the sense of being known a priori and binding on every rational creature, so I would avoid using Kant's terms. I will not call the demands of ecological ethics categorical, but I will not call them hypothetical. They are not hypothetical in Kant's sense; a hypothetical imperative seems to mean for Kant something like, "If x is what is prudential for you, or necessary to secure what you want, do x." Our farmer was saying something like, "Because I believe y, I must do x." This is not a hypothetical imperative, even if it is not independent of the farmer's world view, not something dictated by abstract reason. Some philosophers treat the world view, which gives to the agent the meaning of situations out of which obligation is recognized, as something the agent can change by an act of will. They consider ethically inadequate an obligation that the agent can avoid by changing opinions. An obligation grounded in desires that can be changed is inadequate, but this is not the case with a person's world view. The world view is not changed without some new knowledge. A world view usually works without the agent's awareness of it. Even when the agent can talk about the world view, it is not subject to the agent's whim. Of course, an agent can lie about beliefs, values, and feelings, but it is not the job of ethics to make people honest; its job is to give moral guidance to honest and well-intentioned people.

What is at stake here can be put much more clearly without using the confusing terms *categorical* and *hypothetical.* A key factor is that a genuine moral imperative stands against the agent; in one sense, the agent may not want to do what is required: a desire to do what is right may be the only reason the agent has for doing it. If a person's individual world view includes findings of ecological science as part of what is true about the world, the obligation this person experiences is an obligation to do what will preserve the integrity of the ecosytem and to avoid what will harm it, and this obligation is felt to be morally binding. Obligations such as this are often referred to as overriding; they tend to override other interests and desires. This overridingness is not absolute; sometimes a competing obligation, perhaps a serious legal obligation or another moral obligation, is so strong that it cannot be overridden. Then the moral agent must leave the first duty undone. Still, the bindingness of this first duty manifests itself in a strong sense of dissatisfaction, often with a feeling that something must be done to make amends. This sense of moral demand is central to understanding morality. I believe it is an adequate interpretation of the demandingness is a necessary condition for an adequate approach to morality. The sense of obligation felt in a particular case might be mistaken, but that is another matter, which has been treated in some detail in Chapters Five and Six. Here I am concerned with discerning a moral demand from a nonmoral one. Of the several criteria that I accept for recognizing a demand as a moral one, the nature of demandingness and the nature of universalizability call for the most discussion and argument.

The sense of overriding demand is easier to comprehend than is the notion of universalizability. Different moral philosophers have interpreted universalizability differently, while a few have expressed doubt about finding a meaningful and nontrivial sense of the concept. I think we can find a nontrivial sense of the requirement, even if it cannot do as much as Kant meant for it to do in an ethical system. It can still be a necessary, if not by itself sufficient, condition for a claim being a moral one. The point I must explain is how a contextual moral demand can meaningfully be described as universalizable. What universalization of a moral claim requires, I believe, is that the demand be treated as one that faces anyone in a situation like the one facing the person making a moral decision. This involves the context in which the moral issue has arisen. The principle on which a person acts is not to be thought of abstractly, as though its application could be independent of every specific context.

A universalized demand can be contrasted with a mere personal policy, a guideline for the self alone. Most of us have certain rules for ourselves that we do not expect other people, regardless of the context, to heed. I always tie my tie with the two ends within a quarter inch of each other. It does not bother me at all to see another person with the two ends hardly within hailing distance of each other. I can even deal comfortably with a tie with the wide end shorter than the narrow end, if the tie is on someone else. My personal rule binds me alone, and it cannot be a moral rule; it is just a matter of my psychological quirk. Quirks, tastes, and mere preferences do not need to be universalized. On the other hand, I believe people who live in a house with a yard similar to mine should compost fruit and vegetable peels and scraps. I know many of them do not do this, and I do not publicly denounce them, as I do not publicly denounce other sinners, but I urge them to do right when the opportunity presents itself. I do this because I believe they should compost. Invalids, people without yards, and people who are away from home most of the time are exempt; their situations are not similar to mine.

Let us drop in on the farmer we talked about earlier. Does he universalize his new land ethic? Now that our farmer knows what should be done with the land, he will try to convert others to ways of farming that will preserve the land, but if he is really knowledgeable, he knows that it would be silly to will that farmers in south Florida, and possibly some other places where the weather and growing conditions are different, use the farming methods that can work where he lives. What nonsense it would be for our morally concerned farmer to condemn as immoral the people who several decades ago started what was seen as promising methods of farming, methods that promised to produce large amounts of food and free farm people from back-breaking labor. Some of them must have rejoiced to read that they were part of a green revolution that would wipe out hunger throughout the world. They did not know they were harming the land; perhaps they had never been taught responsibility to the land, which they thought could take care of itself. These were not bad people, but what they did was terrible.

Let us go further. Does being ethical require that this farmer condemn as morally wrong the reluctance of contemporary farmers whose world views do not include his beliefs? Here we must make some careful distinctions. Moral judgments can be assessments of the character, virtue, integrity, and so on, of the person who acts, or they can be judgments of acts themselves. One might judge an act to be right because it is in keeping with an accepted moral rule. The principle of utility, on the other hand, is a different basis for judgment of the act, asking whether it produces the greatest balance of pleasure over pain. Other moral approaches are more interested in the agent's attitude and reasons for acting. Both the teleological approach, which judges an act on the basis of the consequences of the behavior, and the deontological approach, which assesses the moral value of doing the action, have played significant roles in ethics. A complete ethical system might need elements from both approaches. Our problem here is to determine what must be universalizable in an adequate moral system. If qualities of the agent, such as integrity, good intentions, and an effort to take all relevant matters into consideration, are what can be universalized, is that enough to make the approach a moral one? If we place the weight of moral consideration on such traditional concerns as motives and character when we decide that a decision was a morally right one, we need not be claiming that the decision was the best one that could be made in terms of its consequences. Acting from moral considerations, even acting from what Kantian ethics would consider a good will, might not achieve good consequences; the right might not produce the greatest good. To what extent should we require that the greatest good be done for an act to be considered a morally right act? It may not play any part in our moral assessment of the agent; it may play the major role in our assessment of the act itself. What I am going to suggest is that an ethical approach that recognizes the difficulty in knowing what will be the lasting good can accept this uncertainty and acknowledge that morally good people do not always agree, especially in the assessment of prospective acts in respect to their likely consequences. Such a system need not be subjectivistic; it can have fairly specific standards of ethical decision-making. It can universalize certain aspects of morality. It does not make moral judgment simple, and it does not remove all uncertainty.

I believe I have shown that moral demands stand against a person and tend to override desires and interests, and they are universalizable in a significant way. Contextual demands, even those based on a principle that may not be employed in all morally demanding contexts, are overriding and universalizable in the senses I have explained. I believe this analysis satisfies the legitimate concern of ethicists who have hesitated to accept contextual demands into a system of ethics.

Contextual Obligation and Uncertainty

Let us be clear about the kind of situation that will show plainly clear the difference in respect to the element of uncertainty between the ethical principles I propose and other approaches. In many situations our ethical approach will be able to give definitive ethical guidance. In these situations the choice is between something valued and something not valued, or it is between something valued highly amd something valued slightly. At other times the choice is between something valuable that can be secured only now or only in one way, while the competing value can be secured later. In these situations there is no significant disagreement about valued states of affairs that the morally right action should be seeking; there is no need to be uncertain about what to do. It is only when there is no valued consequence of an action that can be seen to rank above all others that the system does not provide definitive guidance. Situations do arise in which the ranking of values does not fall neatly into place; in such a situation, a rightly intentioned person will not know for certain what should be done; two well-intentioned people could disagree about the best course of action to take. Such situations occur with respect to treatment of the natural environment. The morally serious environmentalist makes a decision but might be reluctant to universalize the specific action.

How serious is it that our ecological ethics leave us from time to time in this position? First let us see that every ethical system runs into a problem at this point. Utilitarianism, even rule utilitarianism, faces this difficulty. Richard Brandt saw that the form of rule utilitarianism that he considered credible would require a "remainder rule" to handle cases in which the substantive and procedural rules did not resolve an issue. Both of his two suggested "remainder rules" had to go outside of rule utilitarianism for their principles.[5]

Contemporary Kantianism has not been able to provide a rule to cover cases in which two important duties of perfect obligation call for different courses of conduct. Ecological ethics is probably no more vulnerable because of the lack of a formal criterion for re-solving the hard cases than is any other ethical system. This is why hard cases are recognized as hard cases; they make demands on the ethical system that the system cannot handle without employing a new, often *ad hoc*, stratagem. That such situations might arise is no reason to condemn the ethical system. Life is complex and often unpredictable; an ethical system that can meet the needs of life can hardly be much more simple than life. Of course, a system of ethics that can give definitive answers to all moral questions looks very appealing at first, but then we realize the artificiality that must be necessary to make this possible. A requirement that a system always be able to give definitive answers is so strict as to be crippling. It is not a serious objection to ecological ethics to find that such ethics works better when it does not claim an absoluteness that no other system has succeeded in justifying. It can acknowledge that it is contextual in that, first, its obligations take their meaning and their moral force from their fittingness in specific situations, and, second, a necessary part of a specific situation is the world view of the person who constitutes that situation and must act in it. As a result of recognizing the unavoidable uncertainty that comes with complex situations, especially unprecedented ones, a certain amount of toleration is appropriate. No matter how confident we may be of our ethical principles, we can be uncertain about the way we have seen a situation. We may not be sure our projection of consequences of an action is correct. This admission of uncertainty is not a weakness of environmental ethics; it is one of its strengths. We are less likely to do the wrong thing if we are honest enough not to claim more certainty than we have a right to claim.

I do not think the contextual nature of the obligations of ecological ethics makes the system weak; in fact, it makes the obligations stronger. Being in the context of a lived situation makes the obligation concrete; there is no abstract separation of the agent from the situation in which a choice must be made and no abstract separation of thought and feeling, of mind and body. When the meaning and importance of the situation are understood on the basis of a world view, the agent's previous life experiences are brought into an active role again. The agent is truly engaged as a whole person in a real situation.

The analysis of moral judgment in this and previous chapters has taken account of two aspects, moral evaluation of the agent and evaluation of the acts of the agent. The acts are evaluated in terms of the justice and the consequences of the deeds in question. The evaluation of acts is sometimes difficult and uncertain. It requires an adequate comprehension of what is going on in the world and of the effects of one's actions on the world. We have seen that certain qualities and behavioral patterns of the agent, such as the agent's character and way of making moral judgments, can be universalized more simply. Integrity of motivation and intention, a sincere will to do justice and cause good to all concerned, is one side of the evaluation of the agent. The other side is the agent's use of the best available method in making a moral judgment, one that is not arbitrary and takes into consideration everything that is known to be relevant, one that follows a careful study of the situation and the probable effects of projected actions. It is of great importance that the person act on the basis of an adequate world view. When we see these in a person's character and decision-making procedures, we can say the person has acted morally. When we see a person adopt an arbitrary and self-serving course of conduct, we withhold our approval. There is nothing contradictory in saying that a person acted morally but did an unfortunate act. In spite of the error, we can respect the person's moral integrity.

The evaluation of an act itself is more difficult than the evaluation of the person. It is not enough to be morally good; the approved act must be the correct act. Mistakes do matter when the act itself is judged. Factual considerations must be weighed, and sometimes the needed information is not yet available. World views must be examined, and there is often room for disagreement about the adequacy of a world view. It is difficult to universalize acts themselves when we do not have enough information for making a firm judgment. The element of uncertainty, however, does not justify our refusal to try to make judgments about acts. At times we do have enough information to justify our criticism of a person for making a bad judgment about a situation or about probable consequences. Because the actions of other people can affect us and the world in which we live, we cannot always withhold judgment even when we are uncertain; the fact that we may err does not justify our refraining from judgment when something important is involved. We should be as responsible as we can be, but we cannot avoid taking chances if we want to do the right thing. Only an ethical system that tries to dispense with consideration of the rightness of actions in terms of their actual consequences or their fittingness with respect to their context can avoid these difficulties. An ethical system that does not consider consequences and contexts would be far too abstract to provide significant guidance for the living of our lives.

The well-intentioned person will receive guidance from the ethical approach I have described, and this is what we have a right to expect from ethics. There is nothing to be gained from having a system of absolute moral obligations that are not grounded in our personal world views and germane to the specific contexts of our lives. Behavior must be performed in specific contexts, and thinking about behavior is clearest when the contextual nature of obligation is recognized. Whether two contexts will be seen as sufficiently similar that the same obligations hold in both is a matter calling for judgment, and sometimes the judgment is easy, while at other times it is difficult. The judgment will depend on how the judge constitutes the situation, and the world view of the person who constitutes these situations will play a significant role in how they will be seen. This does not make the interpretation of the situation whimsical; rather, it grounds it in a lived-world context. We need to remember that constitution is not "seeing what you want to see." When the importance of world views is recognized, we need to remember these world views can be examined and criticized. The moral obligation a person feels when the nature of the constituted life-world makes some actions fitting and others not fitting is genuine moral obligation. It is also important that it be a correct moral decision. When an awareness of the land community and the place of the human being in that community is a salient part of the moral agent's world view, the obligations of environmental ethics will be realized, concretely and demandingly. To make the proper choices in response to our awareness of moral responsibility to the biosphere will require that we use our best thought and all the help we can receive from other people who share our sense of obligation. Perhaps no choices we make will affect other lives as much as the decisions we make about the environment, and few choices will require as much sharing of knowledge, moral insight, and feeling about the planet.

Living with Humanistic Holism

The ethical approach I have presented does not provide moral absolutes that can be applied to all situations we might face. It does not offer us certainty in facing moral issues. It does not promise us ease, simplicity, or a way to achieve unanimity among morally concerned people. What does it offer us? What are the advantages of living with humanistic holism as our guide?

First, let us see that moral absolutes have been available at a cost too great for us to consider. We are not ready to sacrifice our freedom or our intellectual integrity for a sense of certainty. We know we live in a world that is not simple biologically, and we live in societies that are not simple. We know we live in a world that faces many difficult problems. We need the courage to face moral issues without false assurances and without oversimplifications. We must know that a morality that will be adequate for our time cannot be a simple bit of lore passed from generation to generation.

Of course, we will not be attracted to an ethical approach that is more difficult than it needs to be, and I am convinced that a pluralistic and contextualistic humanistic holism is not too complex for us to master or too demanding for us to live by.

We also should not accept an ethical approach that is not morally adequate. I believe the approach I have presented is morally adequate. It does not do what some moral philosophers would like for an ethical system to do, but I have argued that achieving those things is unlikely, because no one has fully succeeded in providing such an ideal system, and we have no reason to expect a morality that can give us absolute rules, give us determinate answers to every moral issue, and allow us to feel certain of the rightness of our decisions. More importantly, I have argued that an ideal system of ethics is not necessary, and the imperfect approach I have presented is morally adequate and will be able to guide our decisions.

Let us review the advantages of a contextualistic and pluralistic approach. The necessity of making our judgments in relation to context is not a troublesome weakness of the approach. It forces us to deal with real situations and to recognize the various factors that are relevant to the decision we are to make. We are not able to retreat into abstraction or to make the judgment simpler through reduction. The judgment we make has a much better chance of being correct and complete because we have to make it with focused attention and with awareness of what is involved.

Pluralism offers us a way to realize all the moral value possible in making a judgment. This may be a bit confusing at times, but longing for simplicity is really longing for a sort of moral poverty, poverty of insight, poverty of sensitivity, poverty of value. Pluralism seems dangerous to some moral philosophers, but we need to put this into perspective. Pluralism might increase our chance of making a new mistake, but it will make us less likely to continue making the same old mistakes. It is up to us to avail ourselves of the richness pluralism makes possible and to use our best judgment so we will avoid as many mistakes as we can.

In ethics we do need to be careful, but we also need to explore, reach out, and broaden ourselves.

Humanistic holism is not for the very timid or the very lazy. It is a demanding approach to morality, but it can give us what we need in facing the coming years. Humanistic holism can enable us to be for people and for the planet.

NOTES

Introduction

1. William Blackstone, ed., *Philosophy and Environmental Crisis* (Athens: University of Georgia Press, 1974).
2. Aldo Leopold, *A Sand County Almanac* (London, Oxford, New York: Oxford University Press, 1949). Paul Shepard and Daniel McKinley, eds., *The Subversive Science: Essays Toward an Ecology of Man* (Boston: Houghton Mifflin Co., 1969).
3. Robert Disch, ed., *The Ecological Conscience* (Englewood Cliffs, N.J.: Prentice Hall, 1970) has articles from the late 1960s; Frederick Elder, *Crisis in Eden* (Nashville: Abingdon, 1970); Alan Shields, "Wilderness, Its Meaning and Value," *Southern Journal of Philosophy* 11 (1973): 240; John Passmore, "The Treatment of Animals," *Journal of the History of Ideas* 36 (1975): 195.
4. Lynn White, Jr., "The Historical Roots of Our Ecologic Crisis," *Science* 155 (1967): 1203.
5. Paul W. Taylor, "The Ethics of Respect for Nature," *Environmental Ethics* 10.3 (Fall 1988): 197; J. Baird Callicott, "The Search for an Environmental Ethic," in *Matters of Life and Death*, ed. Tom Regan, 3d ed. (New York: McGraw-Hill, 1993), 353–354.
6. For my earlier treatment of holism and individuals, see Don E. Marietta, Jr., "Environmental Holism and Individuals," *Environmental Ethics* 3.3 (Fall 1981): 251. Reprinted in Susan J. Armstrong and Richard G. Botzler, eds., *Environmental Ethics: Divergence and Convergence* (New York: McGraw-Hill, 1993), 405–410.

Chapter One

1. Arne Naess, "The Shallow and the Deep, Long-Range Ecology Movements: A Summary" (summary of an introductory lecture at the 3rd World Future Research Conference, Bucharest, 3–10 September 1972), *Inquiry* 16 (1973): 95.
2. William F. Baxter, *People or Penguins: The Case for Optimal Pollution* (New York: Columbia University Press, 1974); A. Myrick Freeman III, "The Ethical Basis of the Economic View of the Environment," in *People, Penguins, and Plastic Trees*, ed. Donald VanDeVeer and Christine Pierce (Belmont, Calif.: Wadsworth, 1986), 218–227.

3. Mark Sagoff, "Some Problems with Environmental Economics," *Environmental Ethics* 10.1 (Spring 1988): 55; see also "At the Shrine of Our Lady of Fatima, or Why All Political Questions Are Not Economic," *Arizona Law Review* 23 (1981): 1283; Steven Kelman, "Cost-Benefit Analysis: An Ethical Critique," *Regulation* (Jan. and Feb. 1981): 74.
4. Steven F. Edwards, "In Defense of Environmental Economics," *Environmental Ethics* 9.1 (Spring 1987): 73; Robert Goodland and George Ledec, "Neoclassical Economics and Principles of Sustainable Development," *Ecological Modelling* 38 (1987): 19.
5. Pete Gunter, "The Big Thicket: A Case Study in Attitudes toward Environment," in *Philosophy and Environmental Crisis*, ed. William T. Blackstone (Athens: University of Georgia Press, 1974), 117–137.
6. Roderick Nash, *Wilderness and the American Mind*, rev. ed. (New Haven: Yale University Press, 1967), Chapters 1 and 2.
7. Karen J. Warren, "Feminism and Ecology, Making Connections," *Environmental Ethics* 9.1 (Spring 1987): 6; see also "The Power and the Promise of Ecological Feminism," *Environmental Ethics* 12.2 (Summer 1990): 128.
8. Merlin Stone, *Ancient Mirrors of Womanhood: Our Goddess and Heroine Heritage* (Boston: Beacon Press, 1979).
9. Michael E. Zimmerman, "Feminism, Deep Ecology, and Environmental Ethics," *Environmental Ethics* 9.1 (Spring 1987): 37.
10. Warren, "Making Connections," 3; "The Power and the Promise," 138–145.
11. John Passmore, "The Treatment of Animals," *Journal of the History of Ideas* 36 (1975): 195–218; Aldo Leopold, "The Land Ethic," in *A Sand County Almanac* (London, Oxford, New York: Oxford University Press, 1949), 201–226.
12. For my early treatment of this issue, see Don E. Marietta, Jr., "Religious Models and Ecological Decision Making," *Zygon* 12.2 (1977): 151.
13. Susan Power Bratton, "Christian Ecotheology and the Old Testament," *Environmental Ethics* 6.3 (Fall 1984): 195; Jeanne Kay, "Concepts of Nature in the Hebrew Bible," *Environmental Ethics* 10.4 (Winter 1988): 309.
14. Eugene C. Hargrove, ed., *Religion and Environmental Crisis* (Athens: University of Georgia Press, 1986).
15. Ti-Fu Tuan, "On Treatment of the Environment in Ideal and Actuality," *American Scientist* 58 (1970): 222.

Chapter Two

1. Aldo Leopold, "The Land Ethic," in *A Sand County Almanac* (London, Oxford, New York: Oxford University Press, 1949);590–599; Arne Naess, "The Shallow and the Deep, Long-Range Ecology Movements: A Summary." (Summary of an introductory lecture at the 3rd World Future Re-

search Conference, Bucharest, 3–10 September 1972), *Inquiry* 16 (1973): 95; Theodore Roszak, *Person/Planet: The Creative Disintegration of Industrial Society*. (Garden City, NY: Anchor Press/Doubleday, 1978).

2. See, for example, Donald Worster, "The Ecology of Order and Chaos," *Environmental Review* 14 (1990): 1; Karen J. Warren and Jim Cheney, "Ecosystem Ecology and Metaphysical Ecology: A Case Study," *Environmental Ethics* 15.2 (Summer 1993): 99; and J. Baird Callicott, "On Warren and Cheney's Critique of Callicott's Ecological Metaphysics," *Environmental Ethics* 15.4 (Winter 1993): 373.
3. Leopold, *Almanac*, 204, 216, 224f.
4. For a critical assessment of the notion of equality between humans and nonhumans, see Holmes Rolston III, *Environmental Ethics: Duties to and Values in the Natural World* (Philadelphia: Temple University Press, 1988): 66–72.
5. Naess, "The Shallow," 95–100.
6. Paul W. Taylor, "The Ethics of Respect for Nature," *Environmental Ethics* 3.3 (Fall 1981): 197.
7. For early articles in support of environmental holism, see George S. Sessions, "Anthropocentrism and the Environmental Crisis," *Humboldt Journal of Social Relations* 6.2 (Fall/Winter 1974): 1; William Devall, "Reformist Environmentalism," *Humboldt Journal of Social Relations* 2 (Spring/Summer 1979): 129; John Rodman, "The Liberation of Nature?" *Inquiry* 20 (1977): 83; Pete Gunter, "The Big Thicket: A Case Study in Attitudes toward Environment," in *Philosophy and Environmental Crisis*, ed. William T. Blackstone (Athens: University of Georgia Press, 1974) 117–137; and Richard Routley and Val Routley, "Against the Inevitability of Human Chauvinism," in *Ethics and Problems of the 21st Century*, ed. K. E. Goodpaster and K. M. Sayer (Notre Dame: University of Notre Dame Press, 1979): 36–59.
8. Lewis Thomas, *The Lives of a Cell: Notes of a Biology Watcher* (Toronto, New York, London: Bantam Books/Viking Press, 1974) and *The Medusa and the Snail: More Notes of a Biology Watcher* (Toronto, New York, London: Bantam Books/Viking Press 1979).
9. Frederick Ferré, "Obstacles on the Path to Organismic Ethics: Some Second Thoughts," *Environmental Ethics* 11.3 (Fall 1989): 231.
10. Garrett Hardin, "Living on a Lifeboat," *BioScience* 24.10 (October 1974): 561.
11. William K. Frankena, "Ethics and the Environment," in *Ethics and Problems*, 3ff. William T. Blackstone, "Ethics and Ecology," in *Environmental Crisis*, 16f, 28f; John Passmore, *Man's Responsibility for Nature: Ecological Problems and Western Traditions* (New York: Charles Scribner's Sons, 1974), 186f.
12. Paul B. Sears, "The Steady State: Physical Law and Moral Choice," in *The Subversive Science*, ed. Paul Shepard and Daniel McKinley (Boston: Houghton Mifflin Co., 1969), 395–401.

13. Thomas B. Colwell, Jr., "The Balance of Nature: A Ground for Human Values," in *Main Currents in Modern Thought* 26 (1969): 46.
14. Holmes Rolston III, "Is There an Ecological Ethic?" *Ethics* 85 (1975): 93.
15. R. W. Sperry, "Science and the Problem of Values," *Zygon* 9 (1974): 7; Ian McHarg, "Values, Process, and Form," in *The Fitness of Man's Environment: Smithsonian Annual II* (Washington, D.C.: Smithsonian Institution Press, 1968), 207–27. Reprinted in Robert Disch, ed. *The Ecological Conscience* (Englewood Cliffs, N.J.: Prentice Hall, 1970), 21–36; E. M. Adams, "Ecology and Value Theory," *Southern Journal of Philosophy* 10 (1972): 3.
16. See, e.g., Laurence H. Tribe. "Ways Not to Think about Plastic Trees," in Tribe, Schelling, and Voss, eds., *When Values Conflict*, ed. Lawrence H. Tribe, Corinne S. Schelling, and John Voss (Cambridge, Massachusett: Ballinger, 1976), 61–91; Mark Sagoff, "On Preserving the Natural Environment," *Yale Law Journal* 84 (1979): 205; Eric Katz, "Utilitarianism and Preservation," *Environmental Ethics* 1.4 (Winter 1979): 357; and Tom Regan, "The Nature and Possibility of an Environmental Ethic," *Environmental Ethics* 3.1 (Spring 1981): 19.
17. For essays on values in nature, see the following papers by Holmes Rolston III: "Values in Nature," *Environmental Ethics* 3.2 (Summer 1981): 113; "Are Values in Nature Subjective or Objective?" *Environmental Ethics* 4.2 (Summer 1982): 125; "Values Gone Wild," *Inquiry* 23 (1983): 181; and "Valuing Wildlands," *Environmental Ethics* 7.1 (Spring 1985): 23. See also Rene Dubos, *The Wooing of the Earth* (New York: Charles Scribner's Sons, 1980).
18. Paul Shepard, *The Tender Carnivore and the Sacred Game* (New York: Charles Scribner's Sons, 1973), 258–278; Roszak, *Person/Planet*, 245–268.
19. E.g., Victor Ferkiss, *The Future of Technological Civilization* (New York: Braziller, 1974); William M. Finnin, Jr. and Gerald Alonzo Smith, eds., *The Morality of Scarcity: Limited Resources and Social Policy* (Baton Rouge: Louisiana State University Press, 1979); Hazel Henderson, *Creating Alternative Futures* (New York: Putnam, 1978); Lester R. Brown, *Building a Sustainable Society* (New York: Norton, 1981).
20. Bryan Norton, ed., *The Preservation of Species: The Value of Biological Diversity* (Princeton: Princeton University Press, 1986) and *Why Preserve Natural Variety* (Princeton: Princeton University Press, 1987).
21. Barry Commoner, *The Closing Circle: Nature, Man, and Technology* (New York: Knopf, 1971), 33.
22. Peter Singer, "Animal Liberation," *New York Review of Books* (April 1973): 17–21; Tom Regan, "The Case for Animal Rights," in *Animal Rights and Human Obligations*, ed. Tom Regan 2nd. ed., (Englewood Cliffs, N.J.: Prentice-Hall, 1989), 105–114.
23. Robert W. Loftin, "The Medical Treatment of Wild Animals," *Environmental Ethics* 7.3 (1985): 231; also Loftin, "The Morality of Hunting,"

Environmental Ethics 6.3 (1984): 241; Paul Shepard, *The Tender Carnivore and the Sacred Game* (New York: Scribner, 1973), 258.

24. J. Baird Callicott, "Animal Liberation and Environmental Ethics: Back Together Again," in J. Baird Callicott, *A Defense of the Land Ethic* (Albany: SUNY Press, 1989); Mary Midgley, *Animals and Why They Matter* (Athens, Georgia: University of Georgia Press, 1984), 112–124.
25. Gary Varner's paper is to be published in a volume tentatively titled *Environmental Philosophy and Environmental Activism.*

Chapter Three

1. Aldo Leopold, *A Sand County Almanac* (London, Oxford, New York: Oxford University Press, 1949), 204; Arne Naess, "The Shallow and the Deep, Long-Range Ecology Movements: A Summary." (Summary of an introductory lecture at the 3rd World Future Research Conference, Bucharest, 3–10 September 1972) *Inquiry* 16 (1973): 95; Paul W. Taylor, "The Ethics of Respect for Nature," *Environmental Ethics* 3.3 (Fall 1981): 211–18.
2. William Aiken, "Ethical Issues in Agriculture," in *Earthbound*, ed. Tom Regan (New York: Random House, 1984), 269; Tom Regan, *The Case for Animal Rights* (Berkeley: University of California Press, 1983), 372; Marti Kheel, "The Liberation of Nature: A Circular Affair," *Environmental Ethics* 7.2 (Summer 1985): 135.
3. Eric Katz, "Organicism, Community, and the 'Substitution Problem'," *Environmental Ethics* 7.3 (Fall 1985): 241.
4. Eric Katz, "Is There a Place for Animals in the Moral Consideration of Nature?" *Ethics and Animals* 4 (1983): 75; Mark Sagoff, "Animal Liberation and Environmental Ethics: Bad Marriage, Quick Divorce," *Report from the Center for Philosophy and Public Policy* 4 (1984): 6.
5. Taylor, "The Ethics of Respect," 211–218; Bill Devall and George Sessions, "The Development of Natural Resources and the Integrity of Nature," *Environmental Ethics* 6.4 (Winter 1984): 305.
6. J. Baird Callicott, "Animal Liberation: A Triangular Affair," *Environmental Ethics* 2.4 (Winter 1980): 326.
7. Arne Naess, "The Shallow and the Deep," 95; Bill Devall and George Sessions, "Integrity of Nature," 303, 318; Paul W. Taylor, "Ethics of Respect," 243.
8. J. Baird Callicott, "The Search for An Environmental Ethic," in *Matters of Life and Death*, ed. Tom Regan. 3rd ed. (New York: Random House, 1980), 411; "The Conceptual Foundations of the Land Ethic," in *Companion to A Sand County Almanac*, ed. J. Baird Callicott (Madison: University of Wisconsin Press, 1987), 93f.; and "Animal Liberation and Environmental Ethics: Back Together Again," in J. Baird Callicott, *A Defense of the*

Land Ethic (Albany: State University of New York Press, 1989), 55f. For more detailed explanations of concepts similar to Callicott's nested rings, see Peter S. Wenz, *Environmental Justice* (Albany: State University of New York Press, 1988), Chapter 14, and Christopher D. Stone, *Earth and Other Ethics: The Case for Moral Pluralism* (New York: Harper and Row, 1987), 133–148.

9. Katz, "Organicism," 252.
10. Barry Commoner, *The Closing Circle: Nature, Man, and Technology* (New York: Knopf, 1971), 33–39.
11. For a good and readable overview of the development of ecology, see Donald Worster, *Nature's Economy* (Garden City, NY: Anchor, 1979).
12. Leopold, *A Sand County Almanac*, 216.
13. Commoner, *Closing Circle*, 33f.
14. Paul W. Taylor, "In Defense of Biocentrism," *Environmental Ethics* 5.3 (Fall 1983): 239.
15. See, for example, Donald Worster, "The Ecology of Order and Chaos," *Environmental Review* 14.1–2 (1990): 1; Karen J. Warren and Jim Cheney, "Ecosystem Ecology and Metaphysical Ecology: A Case Study," *Environmental Ethics* 15.2 (Summer 1993): 99; and J. Baird Callicott, "On Warren and Cheney's Critique of Callicott's Ecological Metaphysics," *Environmental Ethics* 15.4 (Winter 1993): 373.
16. Callicott, "Animal Liberation," 324.
17. Peter Miller, "Value as Richness: Toward a Value Theory for the Expanded Naturalism in Environmental Ethics," *Environmental Ethics* 4.2 (Summer 1982): 101.
18. J. Baird Callicott, "On the Intrinsic Value of Nonhuman Species," in *The Preservation of Species: The Value of Biological Diversity*, ed. Bryan G. Norton (Princeton: Princeton University Press, 1986), 138–172; Hume's *Is/Ought* Dichotomy and the Relation of Ecology to Leopold's Land Ethic," *Environmental Ethics* 4.2 (Summer 1982): 163; and "Intrinsic Value, Quantum Theory, and Environmental Ethics," *Environmental Ethics* 7.3 (Fall 1985): 257.
19. Holmes Rolston III, "Values in Nature," *Environmental Ethics* 3.2 (Summer 1981): 113; "Are Values in Nature Subjective or Objective?" *Environmental Ethics* 4.2 (Summer 1982): 125; "Values Gone Wild," *Inquiry* 23 (1983): 181; "Valuing Wildlands," *Environmental Ethics* 7 (Spring 1985): 23; and *Environmental Ethics: Duties to and Values in the Natural World* (Philadelphia: Temple University Press, 1988), 32, 41.
20. Callicott, "Animal Liberation," 326.
21. Robinson Jeffers, *The Double Axe and Other Poems* (New York: Liveright, 1977), *xxi*.
22. Edward MacCurdy, ed. and trans., *The Notebooks of Leonardo da Vinci*, (New York: Reynal and Hitchcock, 1938), vol. 2, 509.
23. Katz, "Organicism," 241.

24. Mary Midgley, *Animals and Why They Matter* (Athens: University of Georgia Press, 1984), 112–124.
25. Callicott, "Animal Liberation and Environmental Ethics," 51–57.
26. Naess, "The Shallow and the Deep," 96–98.

Chapter Four

1. George S. Sessions, "Anthropocentrism and the Environmental Crisis," *Humboldt Journal of Social Relations* 2 (Fall/Winter 1974): 1, 12.
2. Richard Routley and Val Routley, "Against the Inevitability of Human Chauvinism" in *Ethics and Problems of the 21st Century*, ed. K. E. Goodpaster and K. M. Sayer (Notre Dame: University of Notre Dame Press 1979), 36, 37, 56f.
3. Bryan G. Norton, "Environmental Ethics and Weak Anthropocentrism," *Environmental Ethics* 6.2 (Summer 1984): 131–148. Eugene C. Hargrove also remarks on the error of identifying the concept of intrinsic value with nonanthropocentrism, instrumental value with anthropocentrism. See *Foundations of Environmental Ethics* (Englewood Cliffs, NJ: Prentice Hall, 1989): 10f, 124; also "Weak Anthropocentric Value," *The Monist* 75.2 (April 1992): 183.
4. Eugene C. Hargrove argues against the view that anthropocentrism provides no basis for preservation. He says that preservation can be strongly supported through anthropocentric values; in fact the anthropocentric approach is the best way to support preservation of nonliving things, such as caves. See *Foundations of Environmental Ethics*, 124–129, 167.
5. Norton's requirement that preferences be in keeping with an adequate world view is similar to a point I made in a paper in *Environmental Ethics* 4.2 (Summer 1982), in which I held that moral judgments are made in accord with a world view that can be subjected to scrutiny and evaluated on the basis of several criteria. Karen Warren made a similar point. Mark Sagoff ("At the Shrine of Our Lady of Fatima, or Why Political Questions Are Not All Economic," *Arizona Law Review* 23 [1981]: 1283) and others have pointed out the inadequacy of the market concept of value.
6. Are there virtues associated with anthropocentrists and nonanthropocentrists? Geoffrey B. Frasz ("Environmental Virtue Ethics: A New Direction in Environmental Ethics," *Environmental Ethics* 15.3 [Fall 1993]: 259) explores the possibility of relating virtue ethics to environmental ethics and cites the earlier work of Mary Midgley and others.

Chapter Five

1. Aldo Leopold, "The Land Ethic," in *A Sand County Almanac* (London, Oxford, New York: Oxford University Press, 1949), 224f.
2. Holmes Rolston III, "Is There an Ecological Ethic?" *Ethics* 85 (1975): 93.
3. Marietta, Don E., Jr., "The Interrelationship of Ecological Science and Environmental Ethics," *Environmental Ethics* 1.3 (Fall 1979): 195; and "Knowledge and Obligation in Environmental Ethics: a Phenomenological Analysis," *Environmental Ethics* 4.2 (Summer 1982): 153.
4. Kenneth R. Pahel, "Stephen Pepper's Ethical Empiricism and the Myth of Analytic Naturalism," *Southern Journal of Philosophy* 5.1 (Spring 1967): 48.
5. Abraham H. Maslow, "Fusions of Facts and Values," *American Journal of Psychoanalysis* 23 (1963): 120.
6. See Ray Lapley, *Value: A Cooperative Inquiry* (New York: Columbia University Press, 1949), 167–189, and Nicholas Rescher, *Introduction to Value Theory* (Englewood Cliffs, N.J.: Prentice Hall, 1969), 10f.
7. Rolston, "Ecological Ethic," 100f.
8. Arne Naess, "The Shallow and the Deep, Long-Range Ecology Movement: A Summary," *Inquiry* 16 (1973): 98.
9. Several moral philosophers have seen a close and immediate connection between factual knowledge and moral choices. See Maurice Mandelbaum, *The Phenomenology of Moral Experience* (Baltimore: Johns Hopkins University Press, 1969), especially chapters 2, 5, and 6; also John Rawls, "Outline of a Decision Procedure for Ethics," *Philosophical Review* 66 (1957): 177.
10. Many of the findings about perception that I will use come from phenomenological research. A readable textbook on phenomenology is David Stewart and Algis Mickunas, *Exploring Phenomenology* (Chicago: American Library Association, 1993). I am also indebted to findings of gestalt psychologists, especially Abraham Maslow, to whom I refer frequently in several chapters.
11. Marietta, "Knowledge and Obligation," 156.
12. See especially Rudolf Arnheim, *Art and Visual Perception*, (Berkeley: University of California Press, 1974).
13. Irving Thalberg, "Visceral Racism," *The Monist* 56.4 (1972): 43.

Chapter Six

1. E.g., Peter Singer, "The Triviality of the Debate Over 'Is-Ought' and the Definition of 'Moral'," *American Philosophical Quarterly* 10 (1973): 51. Singer describes the 'is/ought' problem as the logical difficulty of bringing together three elements, description, value, and reasons for acting. As he

sees it, justification of a moral position requires two logical steps; first comes the move from the factual 'is', the descriptive element, to value, the knowledge of what is good. The second step is from value to 'ought'. Singer holds that the first part, the relationship between fact and value, and the second part, the relationship between value and obligation, are logically separate matters, and resolving the first does not lead logically to the resolution of the second.

2. This ethical school is associated primarily with Richard M. Hare. See *Freedom and Reason* (New York: Oxford University Press, 1963).
3. Peter Singer says that ethicists who try to combine the advantage of descriptivism (which gives value a grounding in fact) with the advantage of neutralism (which makes value judgments prescribe actions, by definition) slide from one definition of value to another. If Singer is right about this, the slide is an equivocation because it would require the employment of two incompatible definitions of value.
4. E.g., Alan Gewirth sees the connection between moral valuation and action in the fact that purposive action is "conative and dynamic" and follows from a purpose the agent has deemed good, deemed worth the effort. There is for Gewirth no gap between the normative judgment and the reasons for acting. See Alan Gewirth, *Reason and Morality* (Chicago: University of Chicago Press, 1978), Chapters 1 and 2.
5. I have found that the relationship between values and moral obligations can be understood best through a phenomenology of moral action. In phenomenological analysis of moral decision-making, the unification of value and reasons for acting is not to be understood through logical argument. It is found in the phenomenology of the relationship of fact, value, and volition in the constitution of experience. This relationship is most complete in technological and moral decisions. For a clear treatment of this, see Lester Embree, "Some Noetico-Noematic Analyses of Action and Practical Life," in *The Phenomenology of the Noema*, ed. J. J. Drummond and L. Embree (Dordrecht, The Netherlands: Kluwer Academic Publishers, 1992), 198ff.
6. Maurice Mandelbaum, *The Phenomenology of Moral Experience* (Baltimore: Johns Hopkins University Press, 1969), 60–70.
7. Abraham H. Maslow, "Fusions of Facts and Values," *American Journal of Psychoanalysis* 23 (1963): 120, 125–129.
8. Mandelbaum, *The Phenomenology of Moral Experience*, 71.
9. Holmes Rolston III, "Are Values in Nature Subjective or Objective?" *Environmental Ethics* 4.2 (Summer 1982): 150.

Chapter Seven

1. J. Baird Callicott, "On the Intrinsic Value of Nonhuman Species," in *The Preservation of Species: The Value of Biological Diversity*, ed. Bryan G.

Norton (Princeton: Princeton University Press, 1986), 138–172; "Hume's *Is/Ought* Dichotomy and the Relation of Ecology to Leopold's Land Ethic," *Environmental Ethics* 4.2 (Summer 1982): 163; and "Intrinsic Value, Quantum Theory, and Environmental Ethics," *Environmental Ethics* 7.3 (Fall 1985): 257.

2. Holmes Rolston III, "Values in Nature," *Environmental Ethics* 3.2 (Summer 1981): 113; "Are Values in Nature Subjective or Objective? *Environmental Ethics* 4.2 (Summer 1982): 125; "Values Gone Wild," *Inquiry* 23 (1983): 181; "Valuing Wildlands," *Environmental Ethics* 7.1 (Spring 1985): 23; and *Environmental Ethics: Duties to and Values in the Natural World* (Philadelphia: Temple University Press, 1988): 32, 41.
3. J. Baird Callicott, "Intrinsic Value, Quantum Theory, and Environmental Ethics," *Environmental Ethics* 7.3 (Fall 1985): 257.
4. Don E. Marietta, Jr., "Thoughts on the Taxonomy and Semantics of Value Terms," *The Journal of Value Inquiry* 25.1 (1991): 43.
5. Rolston, "Subjective or Objective?" 131f.
6. See especially "Values Gone Wild," "Valuing Wildlands," and *Environmental Ethics: Duties to and Values in the Natural World.*
7. "Elegy Written in a Country Churchyard"
8. See Holmes Rolston III, "Disvalues in Nature," *The Monist* 75.2 (April 1992): 250.
9. Abraham H. Maslow, "Fusions of Facts and Values," *American Journal of Psychoanalysis* 23 (1963): 127–129.
10. See Lester Embree, "Some Noetico-Noematic Analyses of Action and Practical Life," in *The Phenomenology of the Noema*, ed. J. J. Drummond and L. Embree (Dordrecht, The Netherlands: Kluwer Academic Publishers, 1992), especially 198ff.

Chapter Eight

1. Jonathan Harrison, "Utilitarianism, Universalization, and Our Duty to Be Just," *Proceedings of the Aristotelian Society* (1952–1953), 105–34.
2. Marcus G. Singer, *Generalization in Ethics*, (New York: Alfred A. Knopf, 1961).
3. Joseph Fletcher, *Moral Responsibility* (Philadelphia: The Westminster Press, 1967).
4. *Nicomachean Ethics* II.6.
5. J. Baird Callicott, "Animal Liberation and Environmental Ethics: Back Together Again," in *A Defense of the Land Ethic* (Albany: State University of New York Press, 1989).
6. At a conference on Environmental Philosophy and Environmental Activism in May 1993, Gary E. Varner explored grounds for resolving the dispute between defenders of animals and defenders of the environment. The papers

from this conference will be published in a volume of the same name as the conference.

Chapter Nine

1. Bernard Williams, *Ethics and the Limits of Philosophy* (Cambridge: Harvard University Press, 1985); Eugene C. Hargrove, "The Role of Rules in Ethical Decision Making," *Inquiry* 28 (1985): 22, 24, 30; Andrew Brennan, *Thinking About Nature: An Investigation of Nature, Value, and Ecology* (Athens: University of Georgia Press, 1988), 186; Gary E. Varner, "No Holism without Pluralism," *Environmental Ethics* 13.2 (Summer 1991): 175.
2. Peter S. Wenz, *Environmental Justice* (Albany: State University of New York Press, 1988); Christopher D. Stone, *Earth and Other Ethics: The Case for Moral Pluralism* (New York: Harper and Row, 1987) and "Moral Pluralism and the Course of Environmental Ethics," *Environmental Ethics* 10.2 (Summer 1988): 139.
3. Anthony Weston, "Before Environmental Ethics," *Environmental Ethics* 14.4 (Winter 1992): 321.
4. Carol Gilligan, *In a Different Voice* (Cambridge: Harvard University Press, 1982); see also Gilligan et al., ed., *Mapping the Moral Domain: A Contribution of Women's Thinking to Psychological Theory and Education* (Cambridge: Harvard University Press, 1988).
5. Karen J. Warren, "Feminism and Ecology, Making Connections," *Environmental Ethics* 9.1 (Spring 1987): 3; and "The Power and the Promise of Ecological Feminism," *Environmental Ethics* 12.2 (Summer 1990): 125.
6. Warren, "The Power and the Promise," 139.
7. Jim Cheney, "Postmodern Environmental Ethics: Ethics as Bioregional Narrative," *Environmental Ethics* 11.2 (Summer 1989): 117; and "The Neo-Stoicism of Radical Environmentalism." *Environmental Ethics* 11.4 (Winter 1989): 293.
8. See note 2 above.
9. *Ibid.*
10. J. Baird Callicott, "The Case Against Moral Pluralism," *Environmental Ethics* 12.2 (Summer 1990).
11. Callicott, "The Case Against Moral Pluralism," 113, 115, 120f, 121, 123.
12. Anthony Weston, "Comment on Callicott's Case Against Pluralism," *Environmental Ethics* 13.3 (Fall 1991): 283.
13. Callicott seems to accept principle pluralism but is disturbed by lack of a unifying metaphysical foundation. His basis for opposing the approaches of Wenz and Stone is not their use of multiple moral principles but what he sees as lack of metaphysical unity, and his reason for calling them pluralistic is not the same as mine.

14. William K. Frankena, *Ethics*, 2nd ed. (Englewood Cliffs, N.J.: Prentice Hall, 1973), 43f.

Chapter Ten

1. Maurice Mandelbaum, *The Phenomenology of Moral Experience* (Baltimore: Johns Hopkins University Press, 1969), 30–43, 240–244.
2. Thomas Nagel, *Mortal Questions* (Cambridge: Cambridge University Press, 1979), 128–134.
3. William David Ross, *The Right and the Good* (Oxford: Clarendon Press, 1930), 30–32.
4. See Paul W. Taylor, *Respect for Nature: A Theory of Environmental Ethics* (Princeton: Princeton University Press, 1986), 269–273.
5. Donald VanDeVeer, "Interspecific Justice," *Inquiry* 22.1–2 (1979): 55.

Chapter Eleven

1. See Alan Gewirth, "The 'Is-Ought' Problem Resolved," *Proceedings and Addresses of the American Philosophical Association* 48 (1973–1974): 36, 43 for an example of the strictures on morality that I think are detrimental to the practice of morality because we are able to accommodate our morality to them only in very limited and abstract ways. Gewirth argues that moral obligations must be categorical in a sense that will be described more fully later. My main point of disagreement with Gewirth is over the role of the moral agent's personal world view.
2. Philippa Foot, "In Defense of the Hypothetical Imperative," *Philosophic Exchange* 1 (1971): 137; "Morality as a System of Hypothetical Imperatives," *Philosophical Review* 81 (1972): 303; William K. Frankena, "The Philosopher's Attack on Morality," *Philosophy* 49 (1974): 345.
3. See, e.g., Alan Gewirth, *Reason and Morality* (Chicago: University of Chicago Press, 1978), 1, 21, who says that categorical obligations cannot be avoided by "shifting one's inclinations, opinions, or ideals." I agree with him about certain kinds of inclination, but Gewirth seems to rule out aspects of the person's world view that are an important aspect of moral judgment.
4. Gewirth acknowledges that obligations proceed "from within the standpoint of the agent, including the evaluations and right-claims he makes," but he thinks the agent will be logically compelled to make certain judgments, namely to recognize a generic right and obligation to claim freedom and well-being for the self and for others. The reason this moral judgment seems to come inevitably from within the standpoint of the agent seems to

be the high level of generality with which obligations are described (*Reason and Morality*, 135, 158f, 169).

5. Richard B. Brandt, "Toward a Credible Form of Utilitarianism," in *Morality and the Language of Conduct* ed. Hector-Neri Castaneda and George Nakhnikian (Detroit: Wayne State University Press, 1963), 133f.

BIBLIOGRAPHY

Adams, E. M. "Ecology and Value Theory." *Southern Journal of Philosophy* 10 (1972): 3–6.

Aiken, William. "Ethical Issues in Agriculture." In *Earthbound*, edited by Tom Regan. New York: Random House, 1984.

Aristotle. *Nicomachean Ethics* II.6.

Armstrong, Susan J. and Richard G. Botzler. *Environmental Ethics: Divergence and Convergence*. New York: McGraw-Hill, 1993.

Arnheim, Rudolf. *Art and Visual Perception*. Berkeley: University of California Press, 1974.

Baxter, William F. *People or Penguins: The Case for Optimal Pollution*. New York: Columbia University Press, 1974.

Blackstone, William T., ed. *Philosophy and Environmental Crisis*. Athens: University of Georgia Press, 1974.

———. "Ethics and Ecology." In *Philosophy and Environmental Crisis*, edited by William T. Blackstone. Athens: University of Georgia Press, 1974.

Brandt, Richard B. "Toward a Credible Form of Utilitarianism." In *Morality and the Language of Conduct*, edited by Hector-Neri Castaneda and George Nakhnikian. Detroit: Wayne State University Press, 1963.

Bratton, Susan Power. "Christian Ecotheology and the Old Testament." *Environmental Ethics* 6.3 (Fall 1984): 195–209.

Brennan, Andrew. *Thinking About Nature: An Investigation of Nature, Value, and Ecology*. Athens: University of Georgia Press, 1988.

Brown, Lester R. *Building a Sustainable Society*. New York: Norton, 1981.

Callicott, J. Baird. "Animal Liberation: A Triangular Affair." *Environmental Ethics* 2.4 (Winter 1980).

———. "The Search for An Environmental Ethic." In *Matters of Life and Death*. 3d ed., edited by Tom Regan. New York: McGraw-Hill, 1993.

———. "Hume's Is/Ought Dichotomy and the Relation of Ecology to Leopold's Land Ethic." *Environmental Ethics* 4.2 (Summer 1982): 163–174.

———. "Intrinsic Value, Quantum Theory, and Environmental Ethics." *Environmental Ethics* 7.3 (Fall 1985): 257–275.

———. "On the Intrinsic Value of Nonhuman Species." In *The Preservation of Species: The Value of Biological Diversity*, edited by Bryan G. Norton. Princeton: Princeton University Press, 1986.

———. "The Conceptual Foundations of the Land Ethic." In *Companion to A Sand County Almanac*, edited by J. Baird Callicott. Madison: University of Wisconsin Press, 1987.

———. "Animal Liberation and Environmental Ethics: Back Together Again."

In *A Defense of the Land Ethic*. Albany: State University of New York Press, 1989.

———. "The Case Against Moral Pluralism." *Environmental Ethics* 12.2 (Summer 1990): 99–124.

———. "On Warren and Cheney's Critique of Callicott's Ecological Metaphysics." *Environmental Ethics* 15.4 (Winter 1993): 373–374.

Cheney, Jim. "Postmodern Environmental Ethics: Ethics as Bioregional Narrative." *Environmental Ethics* 11.2 (Summer 1989): 117–134.

———. "The Neo-Stoicism of Radical Environmentalism." *Environmental Ethics* 11.4 (Winter 1989): 293–325.

Clement, Roland C. "Watson's Reciprocity of Rights and Duties." *Environmental Ethics* 4.1 (Winter 1979): 353–355.

Colwell, Thomas B., Jr. "The Balance of Nature: A Ground for Human Values." *Main Currents in Modern Thought* 26 (1969): 46–52.

Commoner, Barry. *The Closing Circle: Nature, Man, and Technology*. New York: Knopf, 1971.

Devall, William. "Reformist Environmentalism." *Humboldt Journal of Social Relations* 6.2 (Spring/Summer 1979): 129–158.

Devall, William and George S. Sessions. "The Development of Natural Resources and the Integrity of Nature." *Environmental Ethics* 6.4 (Winter 1984): 293–322.

Disch, Robert, ed. *The Ecological Conscience*. Englewood Cliffs, N.J.: Prentice Hall, 1970.

Drengson, Alan R. "Shifting Paradigms: from the Technocratic to the Person-Planetary." *Environmental Ethics* 2.3 (Fall 1980): 221–240.

Dubos, Rene. "The Wilderness Experience." In *The Wooing of the Earth*. New York: Charles Scribner's, 1980.

Edwards, Steven F. "In Defense of Environmental Economics." *Environmental Ethics* 9.1 (Spring 1987): 73–85.

Elder, Frederick. *Crisis in Eden*. Nashville: Abingdon, 1970.

Embree, Lester. "Some Noetico-Noematic Analyses of Action and Practical Life." In *The Phenomenology of the Noema*, edited by J. J. Drummond and L. Embree. Dordrecht, The Netherlands: Kluwer Academic Publishers, 1992.

Ferkiss, Victor. *The Future of Technological Civilization*. New York: Braziller, 1974.

Ferré, Frederick. "Obstacles on the Path to Organismic Ethics: Some Second Thoughts." *Environmental Ethics* 11.3 (Fall 1989): 231–241.

Finnin, William M., Jr., and Gerald Alonzo Smith, eds. *The Morality of Scarcity: Limited Resources and Social Policy*. Baton Rouge: Louisiana State University Press, 1979.

Fletcher, Joseph. *Moral Responsibility*. Philadelphia: The Westminster Press, 1967.

Foot, Philippa. "In Defense of the Hypothetical Imperative." *Philosophic Exchange* 1 (1971): 137–45.

———. "Morality as a System of Hypothetical Imperatives." *Philosophical Review* 81 (1972): 303–316.

Fox, Warwick. "The Deep Ecology—Ecofeminism Debate." *Environmental Ethics* 11.1 (Spring 1989): 5–25.

Frankena, William K. *Ethics*, 2d ed. Englewood Cliffs, N.J.: Prentice Hall, 1973.

———. "The Philosopher's Attack on Morality." *Philosophy* 49 (1974): 345–356.

———. "Ethics and the Environment." In *Ethics and Problems of the 21st Century*, edited by K. E. Goodpaster and K. M. Sayre. Notre Dame: University of Notre Dame Press, 1979.

Frasz, Geoffrey B. "Environmental Virtue Ethics: A New Direction in Environmental Ethics." *Environmental Ethics* 15.3 (Fall 1993): 259–274.

Freeman, A. Myrick, III. "The Ethical Basis of the Economic View of the Environment." In *People, Penguins, and Plastic Trees*, edited by Donald VanDeVeer and Christine Pierce. Belmont, California: Wadsworth, 1986.

Gewirth, Alan. "The 'Is-Ought' Problem Resolved." *Proceedings and Addresses of the American Philosophical Association* 48 (1973–1974): 34–61.

———. *Reason and Morality*. Chicago: University of Chicago Press, 1978.

Gilligan, Carol. *In a Different Voice*. Cambridge: Harvard University Press, 1982.

——— et al., ed. *Mapping the Moral Domain: A Contribution of Women's Thinking to Psychological Theory and Education*. Cambridge: Harvard University Press, 1988.

Goodland, Robert and George Ledec. "Neoclassical Economics and Principles of Sustainable Development." *Ecological Modelling* 38 (1987): 19–47.

Gunter, Pete. "The Big Thicket: A Case Study in Attitudes toward Environment." In *Philosophy and Environmental Crisis*, edited by William T. Blackstone. Athens: University of Georgia Press, 1974.

Hardin, Garrett. "Living on a Lifeboat." *BioScience* 24, no. 10 (October 1974), 561–568.

Hare, Richard M. *Freedom and Reason*. New York: Oxford University Press, 1963.

Hargrove, Eugene C. "The Role of Rules in Ethical Decision Making." *Inquiry* 28 (1985): 3–42.

———, ed. *Religion and Environmental Crisis*. Athens: University of Georgia Press, 1986.

———. *Foundations of Environmental Ethics*. Englewood Cliffs, N.J.: Prentice Hall, 1989.

———. "Weak Anthropocentric Value." *The Monist* 75, no. 2 (April 1992): 183–207.

Harrison, Jonathan. "Utilitarianism, Universalization, and Our Duty to Be Just." *Proceedings of the Aristotelian Society* (1952–1953): 105–34.

Hartshorne, Charles. "The Rights of the Subhuman World." *Environmental Ethics* 1.1 (Spring 1979): 49–60.

Henderson, Hazel. *Creating Alternative Futures*. New York: Putnam, 1978.

Jeffers, Robinson. *The Double Axe and Other Poems*. New York: Liveright, 1977.

Kant, Immanuel. *Foundations of the Metaphysics of Morals*. Edited by Robert Paul Wolf. Translated by Lewis Beck White. New York: Bobbs-Merrill, 1969.

Katz, Eric. "Utilitarianism and Preservation." *Environmental Ethics* 1.4 (Winter 1979): 357–364.

———. "Is There a Place for Animals in the Moral Consideration of Nature?" *Ethics and Animals* 4 (1983): 74–87.

———. "Organicism, Community, and the 'Substitution Problem'." *Environmental Ethics* 7.3 (Fall 1985): 241–256.

———. "Searching for Intrinsic Value: Pragmatism and Despair in Environmental Ethics." *Environmental Ethics* 9.3 (Fall 1987): 231–241.

———. "Unfair to Foundations? A Reply to Weston." *Environmental Ethics* 10.3 (Fall 1988): 288.

Kay, Jeanne. "Concepts of Nature in the Hebrew Bible." *Environmental Ethics* 10.4 (Winter 1988): 309–327.

Kelman, Steven. "Cost-Benefit Analysis: An Ethical Critique." *Regulation* January and February, 1981: 74–82.

Kheel, Marti. "The Liberation of Nature: A Circular Affair." *Environmental Ethics* 7.2 (Summer 1985): 135–49.

Lapley, Ray. *Value: A Cooperative Inquiry*. New York: Columbia University Press, 1949.

Leonardo da Vinci. *The Notebooks of Leonardo da Vinci*. Edited and translated by Edward MacCurdy. 2 vols. New York: Reynal and Hitchcock, 1938.

Leopold, Aldo. "The Land Ethic." In *A Sand County Almanac*. London, Oxford, New York: Oxford University Press, 1949.

Loftin, Robert W. "The Morality of Hunting." *Environmental Ethics* 6.3 (Winter 1984): 241–250.

———. "The Medical Treatment of Wild Animals." *Environmental Ethics* 7.3 (Winter 1985): 231–239.

Mandelbaum, Maurice. *The Phenomenology of Moral Experience*. Baltimore: Johns Hopkins University Press, 1969.

Marietta, Don E., Jr. "Religious Models and Ecological Decision Making." *Zygon* 12, no. 2 (1977): 151–166.

———. "The Interrelationship of Ecological Science and Environmental Ethics." *Environmental Ethics* 1.3 (Fall 1979): 195–207.

———. "Knowledge and Obligation in Environmental Ethics: A Phenomenological Analysis." *Environmental Ethics* 4.2 (Summer 1982): 153–162.

———. "Environmental Holism and Individuals." *Environmental Ethics* 10.3 (Fall 1988): 251–258. Reprinted in Armstrong, Susan J. and Richard G. Botzler, eds. *Environmental Ethics: Divergence and Convergence*. New York: McGraw-Hill, 1993.

———. "Thoughts on the Taxonomy and Semantics of Value Terms." *The Journal of Value Inquiry* 25, no. 1 (1991): 43–53.

———. "Pluralism in Environmental Ethics." *Topoi* 12 (1993): 69–80.

Martin, John N. "The Concept of the Irreplaceable." *Environmental Ethics* 1.1 (Spring 1979): 31–48.

Maslow, Abraham H. "Fusions of Facts and Values." *American Journal of Psycho analysis* 23 (1963): 120–135.

McHarg, Ian. "Values, Process, and Form," *The Fitness of Man's Environment: Smithsonian Annual II*. Washington, D.C.: Smithsonian Institution Press, 1968. Reprinted in Disch, Robert, ed. *The Ecological Conscience*. Englewood Cliffs, N.J.: Prentice Hall, 1970.

Merleau-Ponty, Maurice. *The Phenomenology of Perception*. Translated by Colin Smith. New York: Humanities Press, 1962.

Midgley, Mary. *Animals and Why They Matter*. Athens: University of Georgia Press, 1984.

Miller, Peter. "Value as Richness: Toward a Value Theory for the Expanded Naturalism in Environmental Ethics." *Environmental Ethics* 4.2 (Summer 1982): 101–114.

Naess, Arne. "The Shallow and the Deep, Long Range Ecology Movement: A Summary" (Summary of an introductory lecture at the 3rd World Future Research Conference, Bucharest, 3–10 September 1972). *Inquiry* 16 (1973): 95–100.

———. "'Man Apart' and Deep Ecology: A Reply to Reed." *Environmental Ethics* 12.2 (Summer 1990): 185–192.

Nagel, Thomas. *Mortal Questions*. Cambridge: Cambridge University Press, 1979.

Nash, Roderick. *Wilderness and the American Mind*. Rev. ed. New Haven: Yale University Press, 1967.

Neihardt, John G. *Black Elk Speaks*. New York: Washington Square Press, 1932.

Norton, Bryan G. "Environmental Ethics and Weak Anthropocentrism." *Environmental Ethics* 6.2 (Summer 1984): 131–148.

———, ed. *The Preservation of Species: The Value of Biological Diversity*. Princeton: Princeton University Press, 1986.

———. *Why Preserve Natural Variety?* Princeton: Princeton University Press, 1987.

———. *The Unity of Environmentalism*. New York: Oxford University Press, 1991.

———. "Epistemology and Environmental Values." *The Monist* 75, no. 2 (April 1992): 208–226.

Pahel, Kenneth R. "Stephen Pepper's Ethical Empiricism and the Myth of Analytic Naturalism." *Southern Journal of Philosophy* 5, no. 1 (Spring 1967): 48–58.

Passmore, John. *Man's Responsibility for Nature: Ecological Problems and Western Traditions*. New York: Charles Scribner's Sons, 1974.

———. "The Treatment of Animals." *Journal of the History of Ideas* 36 (1975): 195–218.

Plumwood, Val. "Ethics and Instrumentalism: A Response to Janna Thompson." *Environmental Ethics* 13.2 (Summer 1991): 139–149.

Povilitis, Anthony J. "On Assigning Rights to Animals and Nature." *Environmental Ethics* 2.1 (Spring 1980): 67–71.

Rawls, John. "Outline of a Decision Procedure for Ethics." *Philosophical Review* 66 (1957): 177–197.

Reed, Peter. "Man Apart: An Alternative to the Self-Realization Approach." *Environmental Ethics* 11.1 (Spring 1989): 53–69.

Regan, Tom. "The Nature and Possibility of an Environmental Ethic." *Environmental Ethics* 3.1 (Spring 1981): 19–34.

———. "The Case for Animal Rights" in Regan, Tom, ed. *Animal Rights and Human Obligations*, 2d. ed. Englewood Cliffs, N.J.: Prentice Hall, 1989.

Rescher, Nicholas. *Introduction to Value Theory*. Englewood Cliffs, N.J.: Prentice Hall, 1969.

Rodman, John. "The Liberation of Nature?" *Inquiry* 20 (1977): 83–131.

Rolston, Holmes, III. "Is There an Ecological Ethic?" *Ethics* 85 (1975): 93–109.

———. "Values in Nature." *Environmental Ethics* 3.2 (Summer 1971): 113–128.

———. "Are Values in Nature Subjective or Objective?" *Environmental Ethics* 4.2 (Summer 1982): 125–151.

———. "Values Gone Wild." *Inquiry* 23 (1983): 181–207.

———. "Valuing Wildlands." *Environmental Ethics* 7.1 (Spring 1985): 23–48.

———. *Philosophy Gone Wild: Essays in Environmental Ethics*. Buffalo: Prometheus Press, 1986.

———. *Environmental Ethics: Duties to and Values in the Natural World*. Philadelphia: Temple University Press, 1988.

———. "Disvalues in Nature." *The Monist* 75.2 (April 1992): 250–278.

Ross, William David. *The Right and the Good*. Oxford: Clarendon Press, 1930.

Roszak, Theodore. *Person/Planet: The Creative Disintegration of Industrial Society*. Garden City, NY: Anchor Press/Doubleday, 1978.

Routley, Richard and Val Routley. "Against the Inevitability of Human Chauvinism." In *Ethics and Problems of the 21st Century*, edited by K. E. Goodpaster and K. M. Sayer. Notre Dame: University of Notre Dame Press, 1979.

Ruether, Rosemary. *New Woman/New Earth: Sexist Idiologies and Human Liberation*. New York: Seabury Press, 1975.

Sagoff, Mark. "On Preserving the Natural Environment." *Yale Law Journal* 84 (1979): 205–267.

———. "At the Shrine of Our Lady of Fatima, or Why Political Questions Are Not All Economic." *Arizona Law Review* 23 (1981): 1283–1298.

———. "Animal Liberation and Environmental Ethics: Bad Marriage, Quick Divorce." *Law Journal* 22.2: 297–307.

———. "Some Problems with Environmental Economics." *Environmental Ethics* 10.1 (Spring 1988): 55–74.

Sears, Paul B. "The Steady State: Physical Law and Moral Choice." In *The Subversive Science*, edited by Paul Shepard and Daniel McKinley. Boston: Houghton Mifflin Co., 1969.

Sessions, George S. "Anthropocentrism and the Environmental Crisis." *Humboldt Journal of Social Relations* 2 (Fall/Winter 1974): 1–12.

Shepard, Paul. *The Tender Carnivore and the Sacred Game*. New York: Charles Scribner's Sons, 1973.

Shepard, Paul and Daniel McKinley, eds. *The Subversive Science: Essays Toward an Ecology of Man*. Boston: Houghton Mifflin Co., 1969.

Shields, Alan. "Wilderness, Its Meaning and Value." *Southern Journal of Philosophy* 11 (1973): 240–253.

Singer, Marcus G. *Generalization in Ethics*. New York: Knopf, 1961.

Singer, Peter. "The Triviality of the Debate Over 'Is-Ought' and the Definition of 'Moral'." *American Philosophical Quarterly* 10 (1973): 51–56.

———. "Animal Liberation." *New York Review of Books*, April 1973, 17–21.

Sperry, R. W. "Science and the Problem of Values." *Zygon* 9 (1974): 7–21.

Stewart, David and Algis Mickunas, *Exploring Phenomenology*. Chicago: American Library Association, 1993.

Stone, Christopher D. *Earth and Other Ethics: The Case for Moral Pluralism*. New York: Harper and Row, 1987.

———. "Moral Pluralism and the Course of Environmental Ethics." *Environmental Ethics* 10.2 (Summer 1988): 139–154.

Stone, Merlin. *Ancient Mirrors of Womanhood: Our Goddess and Heroine Heritage*. Boston: Beacon Press, 1979.

Taylor, Paul W. "The Ethics of Respect for Nature." *Environmental Ethics* 3.3 (Fall 1981): 197–218.

———. "In Defense of Biocentrism." *Environmental Ethics* 5.3 (Fall 1983): 237–243.

———. *Respect for Nature: A Theory of Environmental Ethics*. Princeton: Princeton University Press, 1986.

Thalberg, Irving. "Visceral Racism." *The Monist* 56, no. 4 (1972): 43–63.

Thomas Aquinas. *Summa Theologica*.

Thomas, Lewis. *The Lives of a Cell: Notes of a Biology Watcher*. Toronto, New York, London: Bantam Books/Viking Press, 1974.

———. *The Medusa and the Snail: More Notes of a Biology Watcher*. Toronto, New York, London: Bantam Books/Viking Press, 1979.

Thompson, Janna. "A Refutation of Environmental Ethics." *Environmental Ethics* 12.2 (Summer 1990): 147–160.

Tribe, Laurence H. "Ways Not to Think about Plastic Trees." In *When Values*

Conflict, edited by Laurence H. Tribe, Corinne S. Schelling, and John Voss. Cambridge, Mass.: Ballinger, 1976.

Tuan, Ti-Fu. "On Treatment of the Environment in Ideal and Actuality." *American Scientist* 58 (1970): 222–249.

VanDeVeer, Donald. "Interspecific Justice." *Inquiry* 22, nos. 1–2 (1979): 55–70.

Varner, Gary E. "No Holism without Pluralism." *Environmental Ethics* 13.2 (Summer 1991): 175–179.

Warren, Karen J. "Feminism and Ecology, Making Connections." *Environmental Ethics* 9.1 (Spring 1987): 3–20.

———. "The Power and the Promise of Ecological Feminism." *Environmental Ethics* 12.2 (Summer 1990): 125–146.

Warren, Karen J. and Jim Cheney. "Ecosystem Ecology and Metaphysical Ecology: A Case Study." *Environmental Ethics* 15.2 (Summer 1993): 99–116.

Watson, Richard A. "Self-Consciousness and the Rights of Nonhuman Animals and Nature." *Environmental Ethics* 1.2 (Summer 1979): 99–129.

Watts, Alan. *The Book of the Taboo Against Knowing Who You Are*. New York: Pantheon Books, 1966.

Wenz, Peter S. *Environmental Justice*. Albany: State University of New York Press, 1988.

Weston, Anthony. "Beyond Intrinsic Value: Pragmatism in Environmental Ethics." *Environmental Ethics* 7.4 (Winter 1985): 321–339.

———. "Unfair to Swamps: A Reply to Katz." *Environmental Ethics* 10.3 (Fall 1988): 285–288.

———. "Comment on Callicott's Case Against Pluralism." *Environmental Ethics* 13.3 (Fall 1991): 283–286.

———. "Before Environmental Ethics." *Environmental Ethics* 14.4 (Winter 1992): 321–338.

White, Lynn, Jr. "The Historical Roots of Our Ecologic Crisis." *Science* 155 (1967): 1203–1207.

Williams, Bernard. *Ethics and the Limits of Philosophy*. Cambridge: Harvard University Press, 1985.

Worster, Donald. *Nature's Economy*. Garden City, N.J.: Anchor, 1979.

———. "The Ecology of Order and Chaos." *Environmental Review* 14 (1990): 1–18.

Zimmerman, Michael E. "Feminism, Deep Ecology, and Environmental Ethics." *Environmental Ethics* 9.1 (Spring 1987): 21–44.

INDEX